Hiroaki KOIDÉ

Rethinking nuclear energy

Autopsy of an illusion

Essay translated from Japanese by Asuka BLANCHARD

Golden Nihon Collection / AnimaViva multilingüe

Golden Nihon Collection

dirigée par Jacques KERIGUY

already published:

IKKYÛ. L'impertinence au service de la foi
Sôshô YAMADA, 2012
ISBN 978-99920-1-966-5; ISSN 2305-8455

L'étrange voyage de Confucius au Japon
Olivier ANSART, 2015
ISBN 978-99920-68-07-6; ISSN 2305-8455

Original Japonese title
Genpatsu Zero by Hiroaki KOIDÉ

ISBN 978-4-7790-6093-9 C0295

AnimaViva multilingüe S.L.U.
AD700 Escaldes-Engordany, Principat d'Andorra
www.animaviva-publisher.com

Translation from Japanese: Asuka BLANCHARD
Cover and layout: Alberto JORGE
Printer: Bod GmbH, Norderstedt, Allemagne

ISBN printed book: 978-99920-68-33-5; Legal deposit: British Library 2015
Epub: ISBN 978-99920-68-34-2; Legal deposit: British Library 2015
mobiKindle: ISBN 978-99920-68-35-9; Legal deposit: British Library 2015
ISSN: 2305-8455

To all those who suffer because of nuclear energy

Table of Contents

The endless road of nuclear decommissioning

How to protect ourselves from radioactivity: Q&A

Should we continue using nuclear energy despite everything?

Not to make another mistake

Afterwords

Jacques Keriguy
Director of the Golden Nihon Collection

Who can do justice to Japan without sympathy ;
and how many writers on Japan have a grain of it ?...
But what a horribly difficult thing is to write about Japan !
The effort in itself dries me up.

Lafcadio Hearn, Letter to W. B. Mason,
1st of November 1892.

This collection aims to take a look at the spiritual, social, literary, artistic and cultural anchor of Japan. Each volume of the collection is dedicated to a person who, by their actions, glaring or stealth, by their words, exalted or secretive, by the momentum which moves them, frantic or fragile, has brought a change to the intellectual or sensitive practices of their contemporaries and helped to shape the Japanese imagination. The goal of these bursts of knowledge is to build a framework, from their distant origins, upon which to build the memory of a country buried under a pile of clichés with an elusive view from abroad.

Coming from different research or teaching schools, the authors, either Japanese or Western, all have in common the affinity they share with their subject of study: not content to remain in a time at which it has flourished, they surround the footsteps of its memory to the generations that have preceded until now. To me it seems a way to relate the way of thinking of the Far East and the West, which is still too often considered as opposed, if not irreconcilable.

A collection, the objective of which is to offer its readers reflections on Japanese civilization and culture, cannot remain impartial to the drama of 11 March 2011, which so painfully impacted the country. What further analysis can be added, after the media have spoken so much about the suffering of the victims and denial of the leaders?

How can the truth be addressed when there are so many vested interests, when so many lies built on certainties blur the spirits?

The book of Hiroaki Koide was an immediate success: it was designed for a Japanese audience, but will address all those who, whatever their nationality and language, look for testimonials and indisputable arguments to form an opinion or consolidate an opinion on the risks involved in the production of nuclear energy for civilian or military purposes.

The competence of expertise, accuracy and legitimacy of the observer, the strong impact on the consequences for the population and the environment caused by the disaster justified this choice.

Hiroaki Koide is a well-known scientist; he is also a man of strong convictions reflected in each of his words: Rethinking nuclear energy is therefore an impassioned plea.

Message from the author

Anyone who closes his eyes to the past is blind to the present.
(Extract from the speech by Richard von Weizsäcker,
President of the Federal Republic of Germany, 8 May 1985)

To overseas readers, I would like to thank you for taking this book in your hand.

Human beings have lived through times of wars, and the wars are still going on in several parts of the world. For Western countries and Japan, 70 years have passed since we came out of the Second World War. In Japan, 70 years ago two atomic bombs were dropped, one in Hiroshima and one in Nagasaki. Each city was attacked by only a single bomb, but each city lost the lives of around 100,000 people in a very short period of time. Fortunately (or unfortunately), there were survivors, but they had to live their lives with the painful label "Hibakusha" (A-bomb victims) on their backs. The rest of their lives were miserable because of multiple health problems as well as social discrimination.

Most Japanese people who witnessed the enormous damage and miserable consequences of the atomic bombs were traumatised by the horror. At the same time, because the power of atomic bombs was so overwhelming, we transformed our fear into ambition, as this capacity could be used for a peaceful objective.

After the accident at the Fukushima Daiichi Nuclear Power Plant, which is the topic of this book, many journalists came to interview me. Both Japanese journalists and foreign journalists, including Europeans. There were certain questions that foreign journalists often asked me. One of them was, "Why did Japan start using nuclear power, despite the fact that the Japanese are victims of atomic bombs?" In answer to this question, I responded, "'despite' is not the right word, precisely because we are the only country baptised by atomic bombs, we expected something extraordinary could be produced by nuclear power." As a matter of fact, I was one of them.

The atomic bombs caused us such tragedy; therefore they must be completely abolished. I thought it was Japan's responsibility as the only victim state to show the world that we could use atomic power for good. Then, in 1968 I intentionally chose nuclear engineering at the faculty of engineering as my major at university, and got involved in the field of nuclear study. At the time, it was believed that: nuclear energy was the future energy that would last even after we'd run out of fossil fuel; nuclear power plants would generate electricity with unimaginably low costs; nuclear power plants would never cause an accident as they were the result of the best combined scientific knowledge.

However, as I further studied the field of nuclear energy, I couldn't help but realise that my choice had been a mistake. Uranium that exists beneath the earth's crust is scarcer than fossil fuels. If we calculate how much energy it can produce, it is far less than that of fossil fuels. Nuclear energy generators, even if they work properly, are not cost efficient at all. If we consider the risk of accidents and problems with disposal, the costs would increase astronomically. Finally, a nuclear power plant is a machine. There is no machine that doesn't break. The people who operate the machine are not gods but humans, and humans make mistakes. If that is the case, we might face an enormous accident one day. Above all, the State and the electricity companies who promoted nuclear energy knew the overwhelming dangers that nuclear power plants posed; therefore they imposed the power plants on poor rural areas, instead of constructing them in big cities. When I realised this truth in the autumn of 1970, I started to work towards stopping all the existing nuclear power plants before a catastrophic accident happened. I took a 180° turn and changed the direction of my life. At that time, there were only three nuclear power plants in Japan. Nevertheless, the huge organisation "Nuclear Village" made up of the State, industrial world, academics, and mass media, kept promoting nuclear energy aggressively, and successfully constructed 58 nuclear power plants in the most seismic state in the world.

To my greatest regret, on 11 March 2011, the accident I had anticipated became a reality. The details of the accident at Fukushima are thoroughly described in this book. Four years have passed but we still cannot go near the accident site, and it is not under control at all. As a consequence, many workers are being exposed to radiation every day, and the discharge of contaminated water to the ocean continues. Outside of the power plant, over 100,000 people were forced to leave their homelands. Their peaceful lives were turned upside down, and now they have no place to go back to. Only nuclear power plants can cause such extensive damage. Despite the reality, no one in the Nuclear Village has been held accountable or been arrested.

They have gotten away with a serious crime without being penalised. What they are trying to do today is to re-start nuclear power plants. In Japan, no nuclear power plant has been operating since the accident. They are also promoting the construction of new nuclear power plants and planning to export the nuclear technology to overseas. Why are they still doing such things? Because they learned a lesson from the Fukushima accident: that no matter how much they failed, no matter how much they hurt people, they can get away with it without being punished. The Nuclear Village is a huge system of power. As history shows, a crime of power can only be punished by another system of power.

If another catastrophic nuclear accident happens in the future, a massive amount of radioactive substances will be emitted into the atmosphere. But, even if there isn't any catastrophic accident and the emission of radioactive substances is prevented, the existing radioactive substances, which have already been produced, won't disappear.

In Japan, the first nuclear power plant began operating in 1966. Today, the amount of fission products generated from the nuclear power plant to date is equivalent to 1,300,000 Hiroshima atomic bombs. The first reactor made by mankind began operating in the USA in 1942, with the objective of producing plutonium, the ingredient for atomic bombs. Ever since, research has been undertaken regarding the detoxification of produced radioactive substances, with the great hope that, one day, our science will be able to overcome and eliminate radioactivity. Seventy-two years have passed, but mankind has failed to find any solution. If we cannot eliminate radioactive substances, the only choice left is to isolate them as far as possible from our environment. But the time span of isolation is from one hundred thousand to one million years. The nuclear energy that uses uranium as a resource will last several decades, maybe 100 years for the longest. However, the fission products generated by nuclear energy need to be isolated eternally, even when mankind may not exist.

We should not have made such an ephemeral and foolish choice: pursuing an instant energy while producing toxins, which we can never take care of in the future.

The worst truth is, that the reason Japan chose the path of nuclear power generation was not for energy. In fact, Japan was looking for the capability to produce atomic weapons, by professing it as a peaceful objective. Military needs overlooked the economic and security aspects. Unfortunately, the development of nuclear energy in Japan cannot be stopped easily. However, currently 14 reactors among 58 reactors constructed in Japan have ceased operation or are planned to be stopped and the size of the nuclear industry in Japan will not grow larger than its current size. In Western Europe, 157 reactors existed in the 1990s, the number of which has decreased to 117. It is certain that the transition of history will move towards the end of the nuclear era. However, to not lose sight of a wider perspective of history, to not lose hope, we should acquire the wisdom as mankind in order to abolish nuclear power from the earth as soon as possible.

February 2015

Hiroaki Koidé

Institute of Research on Nuclear Reactors, Kyoto University

Preface

In the 1960s, when I spent my years at junior high school and high school, people said we would run out of combustible fossils, such as carbon and petrol and the era of nuclear energy would take their place. If nuclear energy could be used to generate electricity, our electricity charges would go down to as low as zero. Convinced by such an argument, which had me believe a dream, I majored in nuclear engineering at the faculty of engineering at Tohoku University. However, the more I learned about nuclear energy, the more I realised that the argument was wrong.

Uranium, which is the fuel for nuclear plants, is extremely scarce compared to other resources. The unit cost for generating nuclear energy, if calculated with actual management data, becomes more expensive than hydro or thermal electric generation.

What seemed particularly strange to me when I was still a student was when the Tohoku electric company announced the construction of the Onagawa Nuclear Power Plant in the Oshika peninsula. *"Nuclear energy is so clean and safe that we can even install the plant on rooftops of buildings in the city"*, that's what they said. Then, why don't they build one in Sendai, the city that consumes the most electricity in the region? I found that very strange, so I looked for an answer. The answer, more obvious today, was that nuclear power plants were simply too dangerous. They are so dangerous that big cities with a dense population cannot accept such risks. Therefore, nuclear power plants had always been constructed in sparsely populated areas. If that was the truth, I could no longer dedicate my career to supporting the nuclear industry. Since then, I have been warning of the danger of nuclear power and devoting my research to removing nuclear power plants from where we live.

"Nuclear energy is not expensive", "nuclear energy is safe", "a nuclear accident will never happen". Those are the slogans you have heard and been convinced of in the past. Unfortunately, the accident at Fukushima Daiichi proved the belief was just an illusion. Furthermore, it also proved that *"we will be short of electricity without nuclear energy"* was a lie. In December 2011, Yoshihiko Noda, the former Prime Minister, announced that the accident at Fukushima was under control. However, the situation has been far from under control to date. After more than four years, approximately 150,000 victims are still prevented from returning to their hometowns. At the Fukushima Daiichi Nuclear Power Plant, more than 3,000 workers are working every day to improve the situation in a highly radioactive environment. Discharging contaminated water into the ocean continues. One after another, the serious problems have been revealed. A massive amount of radioactive substances are leaking. It is impossible to control the situation, unless thousands of workers are sacrificed to irradiation.

Faced with such reality, The Prime Minister Shinzo Abe, the leader of the Liberal Democratic Party (LDP)[1] announced a huge lie to the world on 7 September 2013 at the final presentation for promoting the Olympic Games in Tokyo, saying *"The contaminated water is completely blocked." He also said "The situation at Fukushima is under control."* The aim of the Government now, is to restart the nuclear power plants in Japan and export its nuclear technology. The politicians are not the only ones who are promoting the nuclear industry: bureaucrats in the Ministry of Economy, Trade and Industry are doing the same. Despite the serious scale of the accident, they have made policies to emphasise the importance of nuclear energy in the Basic Energy Plan. Moreover, I was shocked to hear that

---

[1] The Liberal Democratic Party has governed Japan since its creation in 1955, with two short interruptions in 1993 and 1994 and during 2009-2012. It takes centre-right position. The current Prime Minister, Shinzo Abe, is recognized as an open nationalist

the documents confirming the policy for promoting nuclear energy were passed around the Ministry in mid-December 2013 (approximately). On the other hand, the Tokyo Electric Power Company (TEPCO), directly responsible for the accident, has also got off scot-free. Despite the fact that it received various benefits from the State due to its financial crisis, its semi-annual financial closing turned out to be positive. It was discovered that a large amount of donations TEPCO promised to stop are still given to the nuclear facilities. Moreover, TEPCO threatened the State with evading the entire payment of the decontamination, which the State offered to pay temporarily. In the end, TEPCO has managed to make the State pay the future decontamination cost from the national budget. Further still, immediately after TEPCO put the subject of re-starting the Kashiwazaki-Kariwa Nuclear Power Plant on the table. I cannot accept such a shameful injustice.

With the hope that the issues of the Fukushima accident are resolved as soon as possible, first, I accept my responsibility for not having been able to prevent the accident from happening. At the same time, with social and legal measures, I pursue the accountability and weight of the sin of those who hid the danger of nuclear energy from the people and promoted it, which ended up causing the accident at the Fukushima Daiichi Nuclear Power Plant.

What is actually happening at the Fukushima Daiichi Nuclear Power Plant?

Contaminated water: not being blocked or controlled

A painful decision: constructing temporary storage tanks

Contaminated water continues to leak at the Fukushima Daiichi NPP (Fukushima Daiichi Nuclear Power Plant). Since early April 2013, when a leak of a massive amount of highly contaminated water was detected coming from the underground water storage structures, TEPCO (the Tokyo Electric Power Company) has stopped using the underground structures. Storage tanks had been set up on the ground level where accumulated contaminated water from underground was progressively transferred. After mid-August, however, new leaks were detected, coming from these ground tanks. In part due to human error, such as water being poured into a tilted tank, but essentially the cause of the leakage was due to the structure of the tanks. Some tanks were installed correctly. However, faced with an emergency situation, many of them were built rapidly with simple techniques: rubber seams inserted between the spaces and steel plates joined with simple bolts. It was obvious that such condition of the tanks caused the leak.

Observing the situation at that time, I must say it could not have been helped, because the level of radiation at the Fukushima Daiichi NPP was too high to work correctly. For example, just patrolling around the tanks can expose workers to radiation. Every single task at the site causes irradiation. Joining steel plates with bolts only takes a short time, but if steel plates need to be welded properly, it takes much longer. The decision for installing temporary tanks was to prevent workers from being exposed to radiation.

Working in a radioactive environment

What is it like to work in a radioactive environment? It is difficult for most people to imagine. I will refer to an extract from the journal written by Norio Hirai, an engineer specialising in nuclear power plants, who worked for a long time at the site as a director. Mr. Hirai contributed a lot at the trial of the Onagawa Nuclear Power Plant. Unfortunately, he passed away in January 1997, at just fifty-seven years old. The direct cause of death was cerebral apoplexy, but he was suffering from liver cancer.

> *"For example, if there is a task to tighten bolts with screws, I will tell you to tighten them diagonally. You are in a radiation controlled area, meaning that you are surrounded by radioactive substances. It's the worst place to be. When you enter the site, you will wear a dosimeter. Everyone has a different time limit because the level of radiation can vary depending on where you are working. It's counted by minutes. But you are not allowed to wear a watch because the watch will be contaminated. So you calculate with your internal body clock. That's how you enter the site.*
>
> *Under such conditions, all you can think of is time. "10 minutes has already passed, 15 minutes has already passed!" while you are tightening up the bolts with the screws. You are afraid when your dosimeter starts beeping (as it has reached the level of irradiation). Newcomers turn blue when they hear their dosimeter beep. You cannot appreciate how it really feels unless you*

experience it. Then, the workers become careless about their job. Then what happens? (...)

Once, while the nuclear plant was in operation, one big screw became loose. An operating nuclear plant generates a tremendous amount of radiation. Thirty workers had to stand by to tighten up that screw. Everyone stood in a line, and ready, set, go! Each one rushed to the screw, which was about 7 metres away. As little as three seconds after reaching the screw, the dosimeter beeps. Time's up! The worker had to go back to the safe area. Some of them just looked for a spanner, and their time was up. Just tightening one screw cost us 160 workers and about 4 million yen.

You may ask, "why not simply stop a nuclear plant for reparation?"

Stopping a nuclear plant for a day means losing hundreds of millions of yen. The electric power companies do their utmost to avoid a nuclear plant being stopped. Radiation is extremely dangerous, but enterprises care more about their profits than human lives."

Extract from the article *"What I witnessed at the nuclear power plant"*, Shukan Gendai (issue 23 April 2011).

This journal was written long before the accident at the Fukushima Daiichi NPP. It may sound like a long time ago, but even today, the workers are facing the same situation, not just at Fukushima but also at all other nuclear power plants.

The desperate fight against contaminated water

After paying such sacrifices, 1,048 cylinder-shaped storage tanks known as flange-types, including the temporary tanks, were set up at the Fukushima Daiichi site at the end of October 2013. Among them, 968 tanks, with a volume of about 333,000 tonnes, contain the contaminated water pumped out from the No.1-No.4 reactors. We were concerned that those tanks might shortly overflow with the water injected into the core as a coolant, as well as the unstoppable seeping underground water.

In the meantime, numerous typhoons hit Japan with continuous heavy rain. At that time, the contaminated water accumulated in the surrounding artificial embankments and the tanks started flowing into the ocean. There was also an error in the water being poured in without the quantity being measured. To make matters worse, they didn't know how to treat the polluted water remaining inside of the barriers of embankments. TEPCO, having no other choice, decided to remove the polluted water from the embankments, and put it in the underground structures they had once given up using due to the series of leakages.

Let me emphasise that it was only meant to be a temporary solution. They have been working on replacing the temporary tanks with permanent ones as well as adding more tanks. In the meantime, the leakage continues.

Besides, the reactor cores need to be constantly cooled down with water. Moreover, infiltrations of underground water add up the volumes.

The amount of contaminated water increases about 400 tonnes per day.

Even if many temporary flange-type tanks were to be constructed rapidly, the capacity of one tank only holds about 450 tonnes. The permanent-type tanks with proper welding require more time to make. But the longer it takes to make tanks, the later it gets to store the contaminated water. Supposing the goal is achieved by sacrificing workers irradiation, soon there will be no space left to place the tanks. Eventually, this project would fall apart.

How did it happen in the first place?

Why the reactors need to be cooled down

On 11 March 2011, the Fukushima Daiichi Nuclear Power Plant was hit by an earthquake and tsunami and lost total electric supply. Ordinary machines just stop when they lose electricity, but not nuclear reactors.

Inside a nuclear power plant, energy is generated by the fission chain reaction of uranium. If there is trouble, it is relatively easy to stop the operation of the fission chain reaction of uranium. When the fission of uranium occurs, it generates 100 million times more radioactive substances than original uranium. The more the nuclear plants operate, the more radioactive substances are accumulated. The heat discharged by the radioactive substances is called "decay heat". This decay heat continues to generate heat even after the reactors have been stopped.

Just to give you an idea, the No.1 reactor, the smallest unit at the Fukushima Daiichi NPP (460 Megawatts) produces about 10 MW immediately after the operation shutdown. This is equivalent to the heat generated from ten thousand household ovens in use at the same time. It is fission products that generate decay heat. Fission products are the sum of about 200 types of radioactive substances with different life spans. Thus, the ones with shorter life spans disappear relatively quickly. One day after the accident, the decay heat diminishes to 10%. The decay heat continues to diminish day by day, but it still generates 5,000KW three days after the reactor was stopped. The heat evaporates 140 litres (equivalent to an ordinary household bathtub) per minute. The No.2 and No.3 reactors were generating 780 MW; meaning they produce higher decay heat. If the heat cannot be removed, the reactor core inevitably melts down. If the reactor core melts, a tremendous amount of radioactive substances will be released into the atmosphere and into the water. The decay heat must be removed somehow. At the time of the accident, however, because of the total blackout, the pumps to inject water as a coolant didn't work. When I saw the critical situation, I advised water be injected immediately. If water was not available, seawater or muddy water should be injected, because water is the best element to cool material down. The Government and TEPCO decided to pump in seawater and send it to the reactor cores. Then they started injecting water by using fire engines. Despite their efforts, all three reactor cores melted down and a tremendous amount of radioactive substances were released to the outside. Since then, to date, they have been injecting water into the core to minimise the amount of radioactivity being released outside.

The injected water did not reach the reactor cores!

When I look back to the time of the seawater injection, there was a strange phenomenon: No matter how much seawater was injected into the reactors, the water levels in the reactors did not change. As a result, they melted down. Why did the water level not rise? The answer to that mystery was revealed two years and nine months later. On 13 December 2013, it was reported that most of the water injected by the fire engines had not reached the core inside of the reactor buildings. There are thousands of pipes inside of the building divided into many channels, which prevented the injected seawater from reaching the reactor core. Each channel has a valve and as one of the valves was not shut, the water flowed into different directions and never reached the reactor core. Apparently, TEPCO knew this fact at the end of March 2011. The reason why they did not disclose it was because *"the information does not contribute to improving the level of security"* (Yomiuri Shinbun). The comment made my jaw drop. At the same time, TEPCO apologised and explained as follows: *"Even if the water injection was sufficient, we are not sure if we could have prevented the meltdown.*

But we probably could have slowed down its progress." (Tokyo Shinbun). I can only agree with the comment that *"the radiation level at the site was so high that it was extremely difficult to close the valves."* (Jiji.com)

Why did the radioactive contaminated water continue to accumulate?

At the beginning of the water injection, TEPCO planned to recycle the used water in the reactor core after purifying it. In other words, the water, once injected into the reactor core, is transformed into contaminated water. If the decontamination system successfully eliminates the cesium, the water can be recycled. If everything had gone as planned, the volume of contaminated water would not have accumulated, but in reality, it did not work out that way. The reactor core, principally speaking, is in the pressure vessel called the reactor pressure vessel. In fact, the reactor pressure vessel had melted and the bottom may have fallen out. Moreover, the reactor containment outside of the reactor pressure vessel, which is designed as a last barrier of defence to confine radioactivity had already been broken. This led to the eruption of hydrogen as well as the radioactivity. The water injected into the bottomless pressure vessel did not stay inside of the reactor. The water flew out into the basement of the reactor building. The building was made out of concrete, a material that cannot store water. Besides, the building would presumably have cracks and holes everywhere due to the severe damage caused from the earthquake. No matter how much water was injected, it leaked out. Therefore, the constant injection of water became necessary in order to cool down the reactor core. The more water that is injected, the more contaminated water is produced: the dilemma continues.

Where would the leaked contaminated water have gone? It would have flown into the basement of the reactor and probably into the turbine building next to it. Moreover, it might have flowed into the underground filled with trenches and tunnels called pits. At the end of March 2011, some 100,000 tonnes of contaminated water was stored at the Fukushima Daiichi NPP site.

A rejected proposal

Anticipating the situation, I made a proposal to TEPCO at the end of March to immediately transport the contaminated water by tankers. TEPCO had already started constructing temporary tanks, rushed by the increasing amount of contaminated water. The temporary tanks would eventually leak but there was not enough time to make solid tanks. A possible waterproof structure that could be used immediately was a tanker. I proposed moving a massive tanker of 100,000 tonnes to the offing of Fukushima and transferring the contaminated water from the nuclear plant to the tanker. Then, the tanker would be transferred to the Kashiwazaki-Kariwa Nuclear Power Plant, which is owned by TEPCO, while the contaminated water would keep accumulating at Fukushima. The Kashiwazaki-Kariwa Nuclear Power Plant is the largest nuclear power plant in the world, and it has an effluent treatment system. That's why I suggested the contaminated water be treated there. Unfortunately, neither the Government nor TEPCO accepted my idea. As a result, the contaminated water has kept accumulating without any solution, and it has now reached around 400,000 tonnes[2]. If they have had accepted my idea, 100,000 tonnes of water could have been pumped out, and the situation would be far simpler than the one they are currently facing. If I think back, I feel a deep sadness and regret that the proposal was not accepted.

Furthermore, since mid-May, I had been suggesting water blockage walls be constructed inside of the basement of the reactor building in order to

---

[2] 555,000 tonnes as at 28 October 2014.

prevent the melted cores from touching the underground water. The cores had already melted down in the No.1 to No.3 reactors. So the question would be, where are the melted cores? TEPCO continued claiming that they were in the reactor containment vessels, unfortunately, no one could go and check if they were still there. It is highly likely that the reactor cores had broken through the bottom of the containment vessels and sank under the ground. If that is the case, they would definitely have made contact with the ground water, and it would be impossible to stop the massive amount of radiation blended in ground water from flowing out to the ocean. TEPCO did not accept this suggestion either. TEPCO, nevertheless, did calculate the cost of constructing the water blockage walls. They estimated 100 billion yen. They figured that it would be too expensive to be accepted at the shareholders meeting, so they turned this idea down. Naoto Kan, the Prime Minister at the time, approved the rejection.

Six months after the accident, in October 2011, TEPCO finally launched the operation to construct the water blockage walls. However, the walls looked like the form of a "U", which means the mountainside is completely open. Not much effect can be expected, as the U form cannot block the ground water from entering. As clear proof, a tremendous amount of highly concentrated radioactive substances have been detected from the ground water of the ocean side in the reactor building.

The problem with frozen soil walls

Finally, the project of constructing frozen soil walls is being put into place with the estimated maintenance budget of 40 billion yen. It's way too late, but I still think they should do it because seepage control walls need to be constructed. Normally, a seepage control wall should be constructed with concrete and steel, but this kind of construction work will take time, which means the workers will be exposed to radiation a long time. If constructing frozen soil walls results in less radiation and can be finished quicker than concrete and steel walls, then maybe it is a better idea for now, although I don't think frozen soil walls last very long.

A technique for freezing soil has been used for tunnel constructions in the past. It's a technique to dig through the underground by freezing a portion of soil. Freezing soil requires electricity. Tunnel construction requires soil to be frozen during certain periods, whereas seepage control walls involve freezing the underground walls with electricity for the length of 1.4 kilometres, 24 hours a day, over future decades. The estimated electricity cost is 3 billion yen per year, which is extremely high. To make matters worse, if there is a power shortage due to an accident or a natural disaster, the frozen soil will melt. The melted walls will cause further serious problems. There is no guarantee that this hypothesis will not occur within the next several decades; therefore, I don't think it is feasible.

Another concern to be addressed is the condition of the ground where the nuclear power plant is standing. The Fukushima Daiichi NPP was constructed on scraped ground a few metres above sea level, where there used to be an elevation of 30-40 metres. Due to the ground water stream from the hill behind the scraped land, the soil is fundamentally vulnerable. Furthermore, the contaminated water, being left for nearly three years, has flown out of the reactor and mixed with the ground water. Much of it has now seeped into the ground. If the freezing soil walls were to be constructed under such conditions to block the ground water, there will be a risk of ground elevation, as well as a risk of contaminated water overflow. Also, if the ground elevation causes liquefaction, the spent fuel storage pool of the No.4 reactor will face a risk of destruction.

After all, even if the freezing soil walls are to be provisionally completed within the next one and half or two years, they will only contribute to slowing down the speed of the contaminated water increase. The only positive thing is that we will be able to gain some time to come up with the next plan.

In short, everything is too late now. The contaminated water has already damaged a large scale of the ground at the site and the ocean. All action was taken too late. Needless to say, it is far from resolving the radioactive contamination. All that can be done is to minimise the negative effect from spreading. These problems have been left unresolved all this time. Today, the situation is critical.

The contaminated water has continued to leak since immediately after the accident

The contaminated water issue is not recent

"The contaminated water leak was found in the underground storage structures". At the beginning of April 2013, the media reported this shocking news as if the problem had come from nowhere. Then, in July, when the media reported *"TEPCO admitted that highly concentrated contaminated water had leaked into the ocean",* public interest increased dramatically. Subsequently, in August, the media announced, *"the tanks on the ground level were also leaking."* In September, it was discovered that *"the water in the underground structure has been leaking for a long time, probably since January last year."* It sounded like no one had any idea about it. Watching the news, my comment was *"Why are they scandalising at this late timing?"*

In fact, the problem of the contaminated water had already existed since immediately after the accident. Since 12 March 2011, when a large amount of seawater began to be injected into the reactor cores, the contaminated water has been stored in the undergrounds of the reactor buildings and the turbine buildings, and the tunnels around them, which connect to the ocean. The water level in those places continued to rise and reached nearly 100,000 tonnes, as I have previously mentioned. At the beginning of April, it was observed that 520 tonnes, 4.7 Giga Becquerel of highly-concentrated mass amounts of contaminated water, cascaded like a waterfall from the crack of the concrete structure called a pit, from which the shore walls were suspended to the port. Some of you may recall the scene. TEPCO, taking the situation very seriously, managed to cover the crack as a result of struggling effort. As soon as the crack was covered, mass media misinformed the situation as *"now the contaminated water leak is blocked"*, and the media has calmed down ever since.

However, the problem has not been resolved at all, because the reactor buildings, the turbine buildings, trenches, pits and wells are all made out of concrete. *A priori,* concrete cracks. There is no concrete structure that never cracks. Without certain processing it cannot store water, besides, the buildings must have had cracks and holes everywhere due to the severe damage caused by the earthquake. It was devastating at the time and still is today, nearly four years later. Therefore, I believe that the contaminated water had already been flowing into the ocean, since the middle of March 2011, immediately after the seawater injection was started. The contaminated water continues to leak. If it has seeped into the underground, it has obviously blended with ground water, and has automatically gone out to the ocean.

Current situation evidenced by the contamination level of Sanriku seaweeds

I have been measuring the radioactive contamination of seaweeds collected at Omoe in Sanriku, (Iwate prefecture). The original objective was to assess the actual level of contamination for the time when the Rokkasho Village Reprocessing Plant would start operating. In fact, the Rokkasho Village Reprocessing Plant is not yet officially operating, but we've been keeping track of annual data every year for the time being. Following the accident at the Fukushima Daiichi NPP, cesium has been detected in Sanriku seaweeds, which is clearly a result of the accident. It is difficult to judge whether the contamination came from the atmosphere or from the ocean current. Yet, it has been confirmed that the ocean far away from the Fukushima Daiichi NPP was contaminated just after the accident.

The measurement was carried out by the gamma ray spectrometry method using a germanium semiconductor detector. This is a method to analyse the radioactive substances (nuclide) that release gamma ray. Nuclides such as strontium and tritium, which release only beta ray but not gamma ray, cannot be measured by this method.

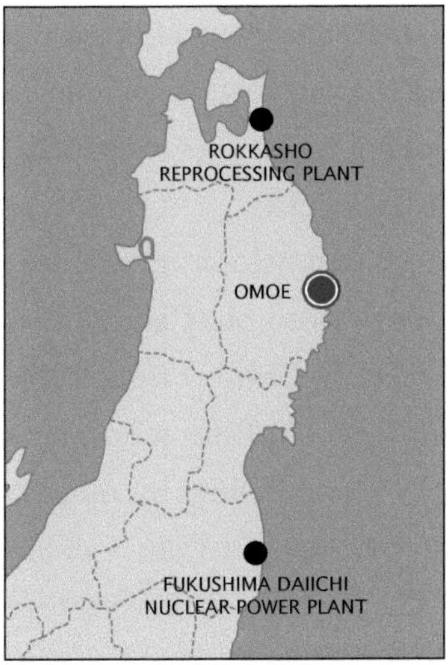

Figure 1: Location of Omoe (Miyako-city, Iwate prefecture)

LIST OF SAMPLES TAKEN	NUCLIDE (HALF-LIFE)		
	K-40 (1.3 billion years)	Cs-137 (30 years)	Cs-134 (2 years)
	Radioactive concentration (Bq/kg)		
Sample taken in April 2006	2 800 ± 5,9	0,19 ± 0,08	ND
Sample taken on 26 March 2007	101 ± 1,3	ND	ND
Sample taken on 26 March 2008	2 190 ± 3,5	ND	ND
Sample taken at the beginning of April 2009	2 650 ± 4,0	0,17 ± 0,04	ND
Sample taken in mid- April 2010 (1)	1 750 ± 3,2	0,21 ± 0,04	ND
Sample taken in mid- April 2010 (2)	1 930 ± 3,5	0,26 ± 0,04	ND
Sample taken in mid- April 2011	1 600 ± 7,7	2,36 ± 0,13	1,75 ± 0,09
Sample taken in mid- April 2012 (1)	1 310 ± 3,4	0,33 ± 0,05	0,14 ± 0,04
Sample taken in mid- April 2012 (2)	1 220 ± 4,1	0,31 ± 0,04	0,098 ± 0,03
Sample taken in mid- April 2013	2 360 ± 4,2	0,85 ± 0,06	0,35 ± 0,05

Table 1: The coast of Sanriku was contaminated immediately after 11 March 2011

Table 1 shows, among the measurable nuclides that emit gamma ray, the result for potassium 40 (K-40), cesium 137 (Cs-137), and cesium 134 (Cs-134). Potassium 40 is one of the typical nuclides found in nature. Cesium 137 is a typical artificial radionuclide, in other words, a fission product made of uranium fission in the nuclear reactor. Its half-life is thirty years. Half-life is a duration required for each radioactive nuclide to reduce the level of radiation by half. The other cesium 134 is also an artificial radionuclide, but its half-life is much shorter (two years).

Thus, if we look at the radioactive contamination measurement in this table, the important data is the amount of cesium. Cesium 137, with its half-life of thirty years, has been detected worldwide through test samples. It was released into the atmosphere as a result of the atmospheric nuclear testing which took place during the 1950s and 1960s by America and the Soviet Union, as well as the Chernobyl accident in 1986. Cesium 134, on the other hand, was not produced by the atmospheric nuclear testing. Even the cesium 134 released after the Chernobyl accident has already reduced and disappeared because its half-life is two years.

If cesium 134 is detected today, there is no doubt that it was caused by the accident at the Fukushima Daiichi NPP.

Bearing that in mind, please take a look at Table 1. Since mid-April 2011, cesium 134 has been constantly detected. Cesium 137 was also detected ten times more in 2011 compared to the concentration before the accident.

I would now like to draw your attention to Figure 2. The figure shows the evolution of cesium 137 and potassium 40 in the seaweeds. These two radioactive nuclides already existed before the accident at the Fukushima Daiichi NPP. The proportion of cesium 137 was a ten thousandth of that of potassium 40. After the accident, the proportion evolved to over one thousandth of potassium 40 and the high ratio has remained ever since.

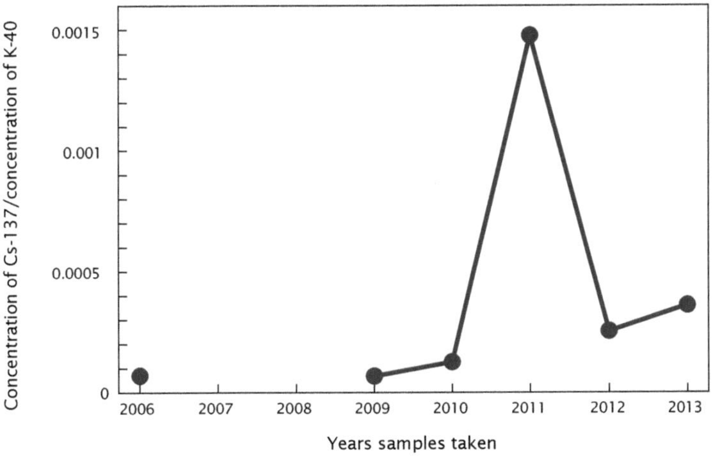

Figure 2: Proportion of Cs-137 and K-40 in the seaweeds

As we can see in the result, ocean contamination is far from under control.

It is clear that the impact of the accident at the Nuclear Plant is extremely serious. Having said that, what is happening to the seaweeds in Sanriku? If you look again at Figure 1 (page 33), cesium 134, because of its half-life of two years, has progressively reduced since immediately after the accident. It was about 70-80% of cesium 137 after the accident, but it has now reduced to less than 50%. As far as observing the result of the measurement is concerned, we can say that the irradiation caused by eating the seaweeds of Sanriku would not be high. The drastic elevation of the quantity in 2013 can be explained not by the accident at Fukushima, but by the level of dehydration of sample seaweeds, as the level of potassium 40 also increased by double that year.

Nuclear power plants pollute the ocean, even without an accident

Let us detract from the accident at Fukushima for now, and I will explain some data that substantiates the fact that a nuclear power plant is an institution that cannot help releasing radioactive substances into the atmosphere.

This is proven by the result of the data researched for the level of ocean contamination around the Ikata Nuclear Power Plant, located in the Ikata village, Nishiuwa district (Ehime prefecture), owned by the Shikoku Electric Power Company. This study is based on the sea-bottom soil collected by the members of a young local researching group dealing with the pollution problems of Isozu. The study started in January 1979 and continues to date, (currently for thirty-five years).[3] Each time, the sample of sea-bottom soil is extracted at four points, shown in Figure 3. The soil is then poured into a plastic container with seawater, and dried while stirring. At the end, the large grains accumulate at the bottom and muddy substances cover the top and eventually, it dries up like skin. The analysis is done both to the grains and to the skin parts, and it appears

[3] The study finished in May 2014.

that the results show a similar tendency. I will introduce you to the result of the analysis on grains, of which we have collected a sufficient and stable amount of samples.

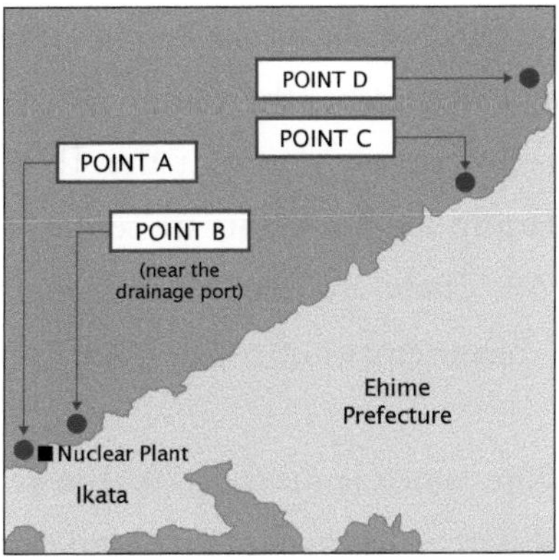

Figure 3: Sample location points around the Ikata Nuclear Power Plant

The method of analysis is identical to the measurement of seaweed contamination in Sanriku: with gamma ray spectrometry using a germanium semiconductor detector. That is to say, radioactive nuclides such as strontium and tritium, that only emit beta ray but not gamma ray, cannot be measured. The natural radionuclides measured are potassium 40 (K-40), and bismuth 214 of the uranium family (U-Series), actinium 228 of the thorium family (Th-Series). Those names are probably not familiar to you. To summarise, all of them are the nuclides that exist in nature. The artificial radionuclides are cesium 137 (Cs-137) and cobalt 60 (Co-60). Cesium 137, as explained previously, can still be found today due to the atmospheric nuclear testing and the accident at the Chernobyl nuclear power plant. As for cobalt 60, on the other hand, because of its relatively short half-life (five years), there should be almost no influence left from past contamination. That means, if cobalt 60 is detected, it has been released from nuclear facilities.

Figure 4-7 (page 39-41) show the result of the analysis of sea-bottom soil from four points. Point A and B are the results of analysis collected near the nuclear power plant, and point B is situated beside drainage.

Let me explain how to look at the tables as they are not easy to understand if you are not familiar with them. First of all, the white dotted horizontal line in the upper part of the table represents the amount of potassium 40. The upside-down triangles, just below, represent the Thorium family, the triangles are the uranium family, the black dots are cesium 137, and the black diamonds are cobalt 60. There are also white diamonds. These are the result of an incomplete measurement of cobalt 60, therefore noted as "doubtful".

The samples have been taken 57 times since January 1979. The values for each nuclide are represented by different marks and recorded on the tables. The horizontal axis indicates years and the vertical axis indicates the quantity of radioactivity per gram. The unit used is Picocuries (pCi). One Picocurie equals 10^{-12}. Currently the unit of radioactivity is Becquerel (Bq); one Picocurie is equivalent to 0.037 Becquerel.

Now, let me explain the result of the analysis. First of all, the values of the three natural radionuclides and the cesium 137 on the upper side of the table are basically quite stable. The cobalt 60, in which we are particularly interested, is not detected at point C or D located far from the nuclear plant, neither at point A, where only small amounts were found at the beginning of the research. At point B, on the contrary, cobalt 60 has been detected much more frequently. In short, the Ikata Nuclear Power Plant has been constantly releasing contaminated water including cobalt 60. Interestingly, sometimes cobalt 60 was not detected at point B. This is due to the border detection limit of the device. If the quantity is below the border, it cannot be measured even if it exists. Since around 1992, cobalt 60 has not been detected because it was below the border.

When the new detector took over, the value of cobalt 60 was detected again. Nevertheless, since the result of the measurement turned out to be "doubtful" in the 51st sample in 1995, it has remained below the border.

In any case, what can be said here is that the contamination of cobalt 60 has gradually been diminishing over time. This is probably because the operation control of the nuclear plant has become more and more strict. We announced the result of the research in 1982. Immediately after the announcement, the official data published by the Shikoku Electric Power Company, which included the result of releasing radioactive effluent liquid except tritium, has fallen below the border of the detection limit. There is no doubt that our campaign had put some pressure on the Shikoku Electric Power Company to react.

This is obviously good news. However, from a technical point of view, it is impossible to reduce the release of radioactive substances to zero. If a more precise investigation were to be undertaken, we would find the trace of radioactive effluent liquid.

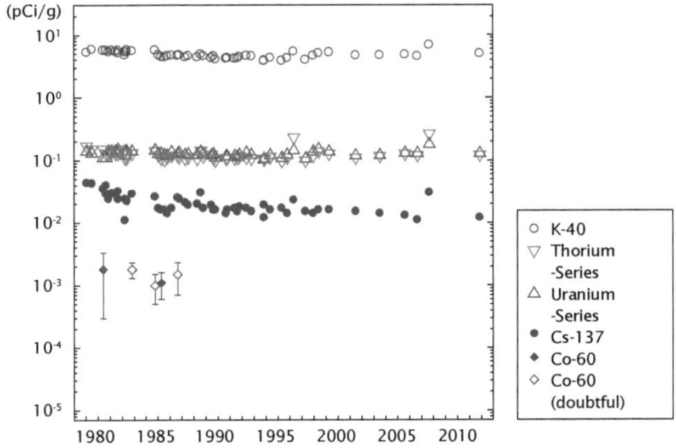

Figure 4: Result of measurement at point A

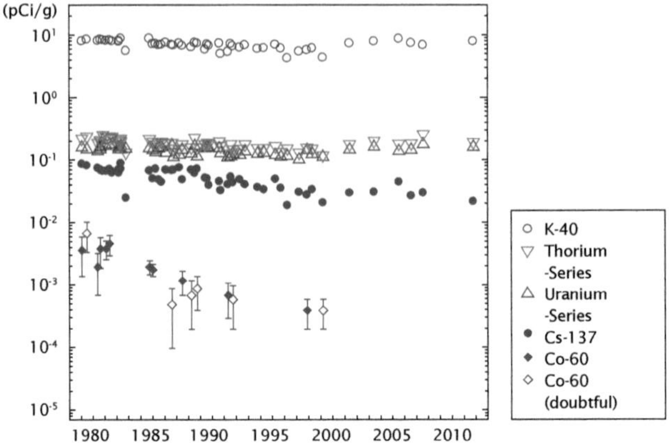

Figure 5: Result of measurement at point B

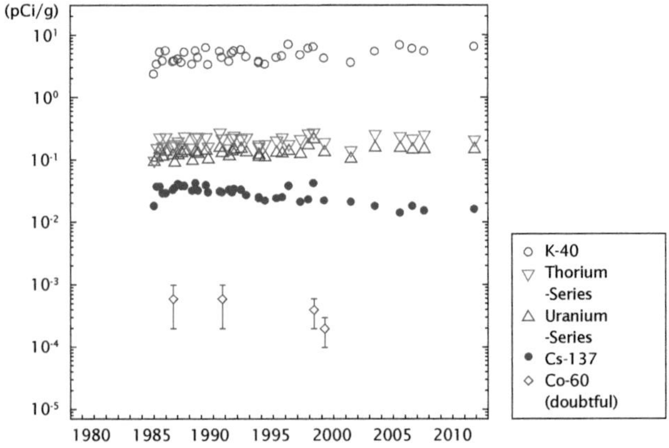

Figure 6: Result of measurement at point C

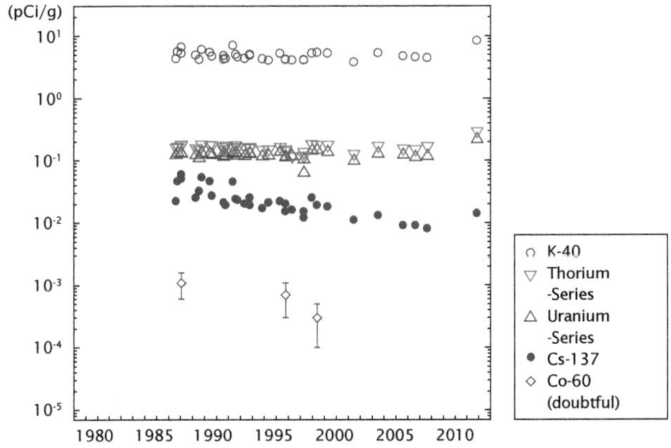

Figure 7: Result of measurement at point D

Cooling with water: a technique reaching its limit

What is tritium, which cannot be filtered even by ALPS?

As explained earlier, the Ikata Nuclear Power Plant seems to control the contaminated water strictly. Nevertheless, it is not possible to get rid of tritium. TEPCO has also been planning to eliminate the radioactive substances from the contaminated water, and release the purified water into the ocean. The device expected to work for this objective is called ALPS (Advanced Liquid Processing System). ALPS is capable of eliminating 62 out of 63 radioactive substances from the existing contaminated water. The only substance left is tritium as every effluent treatment system is useless against tritium.

Tritium is also called Hydrogen 3. It means that it is hydrogen, which is three times heavier than normal hydrogen, but it is still hydrogen. Hydrogen, once released into the atmosphere, transforms into water. To summarise, because the tritium has metamorphosed into water, it cannot be removed even after completely eliminating all the other polluting substances. Radioactivity released from tritium is a very low energy called beta ray, therefore, its toxicity is not so high. For that reason, there is evidence that nuclear power plants, including the Ikata Nuclear Power Plant, have been releasing tritium into the ocean, by justifying it as *"every country is doing it"* and *"no damage to health reported"*. However, it's a false argument to claim that there has been no damage to health to date. No causal relationship was found in the past only because water was being released covertly. Tritium can be a great threat to living beings that depend on water. The humans who drink water can be internally irradiated. The clouds formed by the evaporation of seawater drop rain and pollute lands. Obviously, the damage to animals and plants in general, crops and farms cannot be neglected. Of course, humans can also be irradiated through those living beings by eating them. It is unacceptable to let artificial radioactive substances be released into the ocean, even if its toxicity level is low.

An inevitable choice despite everything: releasing the contaminated water into the ocean

In early November 2013, it was announced that the total volume of contaminated water stored in the tanks inside the Fukushima Daiichi Nuclear Power Plant was 350,000 tonnes.[4]

But that was not the only contaminated water at the site. Estimating the water which has not yet being pumped out and accumulated water in underground tunnels and elsewhere, I assume the total amount will be about 430,000 tonnes. On top of that, 400 tonnes of groundwater would flow into the underground and be added to every day. The contaminated water overflows everywhere at the site.

---

[4] Treated contaminated water (after the elimination of cesium) is 555,000 tonnes and the tank storage volume is about 622,000 tonnes (as at 28 October 2014).

In November, ALPS, which was not functioning from the middle to the end of the month, finally started a test-operation for all three lines. The volume of water that can be treated in a full operation is an average of 500 tonnes per day. It is not sufficient at all and TEPCO reinforced the ALPS systems. The Ministry of Economy, Trade and Industry developed a new ALPS model. Finally, it became possible to treat 2,000 tonnes of contaminated water per day.

They are aiming to finish the contaminated water treatment at the end of March 2015.[5] Personally, I hope the project goes well and the contaminated water disappears as soon as possible. In reality, however, it does not look like it is running smoothly.

To date, ALPS has already had many functional problems: leaks were found here and there, the rubbers that were left during the operation got clogged. Ten days after it started running, chemical products started leaking which forced the ALPS to be stopped. Since the operational work is done in the highly contaminated area, those errors cannot be avoided. ALPS is not a ready-made machine assembled somewhere else, but each part is delivered separately and needs to be put together at the site. Unfortunately, as I explained previously, it is not acceptable to work at the site by taking sufficient time in order to make a reliable machine. For example, if time is running out when connecting pipes by welding them, they will be connected with temporary bands. If the steel plates and steel pipes are too complicated, they will be finished with temporary hoses, otherwise, the irradiation dose rises quickly. Under such conditions, leaks and malfunctions are not at all surprising. It is therefore understandable that the treatment of contaminated water does not work as it's supposed

5 According to TEPCO's report released on 26 May 2015, almost all stored water at Fukushima Daiichi NPP that was contaminated with radioactivity has been treated to remove strontium, and in some cases even more nuclides.

to due to frequent interruptions. In addition, if the treatment does not advance, the more contaminated water will be accumulated and then more tanks will need to be constructed. TEPCO says they will make tanks of up to 800,000 tonnes, but the remaining lots onsite for extension are limited. One day, there will be no choice but to release it into the ocean.

From a fishermen's point of view, this is outrageous. I regret to inform them of such bad news, but there is no other choice. What we are fighting against today is something inextinguishable by human hands no matter what we do. The only thing that can be done is to prevent the leaks. While the only solution is to inject water into the reactor cores, the contaminated water continues to accumulate. The treatment cannot catch up with its speed, and there is no place to store it. When the contaminated water overflows, it has to be diluted as much as possible and dumped into the ocean. The worst news is that, this approach needs to be done before long: we are already at this point.

Think of a counter plan by assembling national and international knowledge

The situation is complex and apparently there is no solution, nevertheless, we must cool down the reactor cores. At the same time, we have to think of a plan to prevent the contaminated water from being released into the ocean. If we inject more water, the more contaminated water will be accumulated. Immediately after the accident, I myself made remarks that any water, even seawater or muddy water, should be injected towards the reactor cores. However, currently, I have begun to think that we should give up on this approach.

Four years on from the accident, some of the radioactive substances have already diminished and the decay heat has reduced to several hundredths compared to immediately after the accident. I think it is possible to cool down the melted reactor cores without depending on water. For the materials that may be able to replace water, I suggest using metal that has a low melting point such as tin, lead, and bismuth. If these metals can reach the reactor cores, they will melt and mix with the reactor cores. By growing the mass, logically the decay heat will decrease, and one day, they will reach the end of dissolution. If they reach this point, they will cool down naturally with the atmosphere. This idea of cooling them with metals, which have low melting points, was advised by Europe after the accident. Recently, other people in Japan have begun to give me such advice too.

Yet, I am not sure if the metal can reach the cores in a suitable condition while the cores have already melted, and we don't even know where they are and what kind of condition they are in. TEPCO once announced that the cores remained in the reactor vessels. As no one can open the cover of the reactor vessels to check if they are there, this hypothesis cannot be confirmed. It may be possible the melted reactor cores have already broken the bottoms of the reactor vessels and fallen out underneath. The Government and TEPCO had an image of the melted reactor cores sitting at the bottom like lumps. But my image is that they have in fact broken into small pieces and scattered everywhere. Besides, melted means that they are in liquid form. That means, they are not staying still in one spot, but may be moving in the pulpy texture called sludge and slurry. It is hard to judge which image is correct unless we can actually see them with our own eyes.

In any case, if metal is used as a coolant, it has to be injected with water by being crushed into small particles. As the current pumps and pipes are curved and bent here and there, the metal would probably get clogged. If that happens, we cannot do anything about it. So we have to seriously examine many issues such as: what kind of pipes should be used? How big should the crushed metal be? How strong should the water current be?

In order to make this approach become reality, the knowledge of hydraulic engineering is needed, and ideas of many other specialists must be integrated.

If sending metal is not feasible, there is also a plan to inject liquid nitrogen and freeze the entire water stored in the reactor vessels. If 5m^3 liquid nitrogen can be injected per hour, it is probably possible. Otherwise, air-cooling such as the ones used for a motorbike engine is a possibility too. All the ideas have no guarantees for success. We are facing an unimaginable reality that humanity has never experienced in the past, and no one can be sure of what to do. For this reason, we should put our knowledge and technology together from all parts of the world, discuss the best solutions and move forward. Unfortunately, the Government and TEPCO have neither the intention nor the system for that.

Tokyo, a radiation controlled area

Current status of the radioactive contamination caused by the accident at Fukushima

Cesium 137: the most dangerous material to mankind

The fission of uranium generates about 200 kinds of fission products (radioactive substances). According to a Government report to IAEA, the international organisation to promote the nuclear industry, three months after the accident, the radioactive substances released into the atmosphere after the accident at the Fukushima Daiichi NPP were: rare gas, iodine, cesium, strontium and plutonium (Table 2). Among them, xenon 133 (1.1×10^{19}) and iodine 131 (1.6×10^{17}) were the materials most released into the atmosphere. Their half-lives are short: 5.2 days for xenon 133 and 8 days for iodine 131, and the other iodine also has a short half-life, between 2-3 hours to within one day. Thus, those materials are not of concern as it is too late and nothing can be done with regards to the irradiation in the future. On the other hand, strontium and plutonium have long life spans and their toxicity is high, but the quantity released into the atmosphere is very small. We therefore don't have to worry too much concerning the damage to soil which leads to vegetable and meat contamination. One exception is strontium, as highly concentrated strontium has been detected inside of the site and we should be careful about marine products.

Therefore, the substance that concerns us the most for the moment is cesium. Although there are two kinds of cesium; cesium 137, which has a longer half-life, is the one that should be taken into consideration. I believe that cesium 137 is the most dangerous radionuclide to mankind.

(unit: Bq)

		Half life	Reactor no. 1	Reactor no. 2	Reactor no. 3	Total
Rare gas	Xe-133	5.2 days	3.4×10^{18}	3.5×10^{18}	4.4×10^{18}	1.1×10^{19}
Iodine	I-131	8 days	1.2×10^{16}	1.4×10^{17}	7×10^{15}	1.6×10^{17}
	I-132	2.3 hours	1.3×10^{13}	6.7×10^{6}	3.7×10^{10}	1.3×10^{13}
	I-133	20.8 hours	1.2×10^{16}	2.6×10^{16}	4.2×10^{15}	4.2×10^{16}
	I-135	6.6 hours	2×10^{15}	7.4×10^{13}	1.9×10^{14}	2.3×10^{15}
Cesium	Cs-134	2.1 years	7.1×10^{14}	1.6×10^{16}	8.2×10^{14}	1.8×10^{16}
	Cs-137	30 years	5.9×10^{14}	1.4×10^{16}	7.1×10^{14}	1.5×10^{16}
Strontium	Sr-89	50.5 days	8.2×10^{13}	6.8×10^{14}	1.2×10^{15}	2×10^{15}
	Sr-90	29.1 years	6.1×10^{12}	4.8×10^{13}	8.5×10^{13}	1.4×10^{14}
Plutonium	Pu-238	87.7 years	5.8×10^{8}	1.8×10^{10}	2.5×10^{8}	1.9×10^{10}
	Pu-239	24,065 years	8.6×10^{7}	3.1×10^{9}	4×10^{7}	3.2×10^{9}
	Pu-240	6,537 ans	8.8×10^{7}	3×10^{9}	4×10^{7}	3.2×10^{9}
	Pu-241	14.4 ans	3.5×10^{10}	1.2×10^{12}	1.6×10^{10}	1.2×10^{12}

Source: Based on the report by the Japanese Government prepared for the ministerial conference of the IAEA on nuclear safety (revised 20 October, 2011)

Table 2: The estimated level of released radioactive substances reported by the Japanese Government to IAEA

Released cesium: equivalent to 168 Hiroshima bombs?

Let's examine Table 2 to see how much of this dangerous cesium 137 was emitted from the Fukushima Daiichi NPP. Please take a look at Figure 8. The numbers may not give you enough understanding about the quantity differences, so I have made a diagram to show you the actual volumes. You can ignore the numbers and just compare the size of the two images.

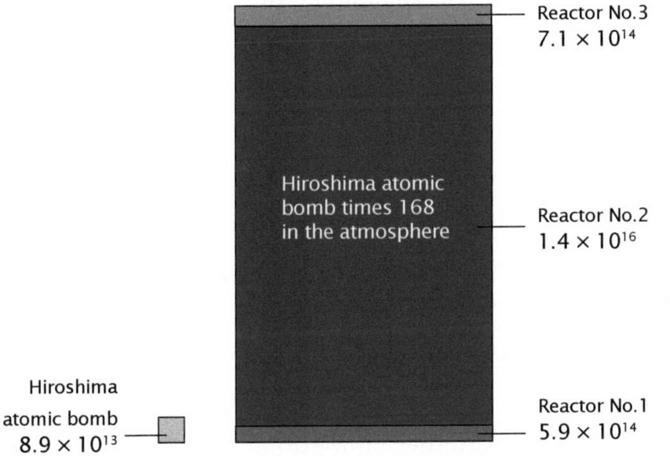

Source: Based on the report by the Japanese Government prepared for the ministerial conference of the IAEA on nuclear safety

Figure 8: Comparison of the quantity of cesium 137 in the atmosphere (Becquerel)

First, I drew a small cube in the lower left corner. That is the volume of the cesium 137 released into the atmosphere by a Hiroshima bomb. If we look at the cube on the right, that is the volume of cesium released by the accident at the Fukushima Daiichi NPP. The No.1 reactor alone emitted 6 or 7 of the Hiroshima bombs; and the No.2 reactor, which emitted a mass amount, shows the volume of countless cubes of Hiroshima bombs. Then the No.3 reactor, although a much smaller volume than the No.2 reactor, also emitted cesium 137. The total volume is equivalent to 168 Hiroshima bombs.

This is already a shocking volume, but because this data is disclosed by the Government, I question the credibility of this information.

Because the Government is the one who convinced us that nuclear power plants were absolutely safe, and thus gave permission to construct 58 plants in this small country. It is not a question of "taking responsibility" but it should be called a "crime". There is no reason for criminals to admit to the full extent of their crime. In my opinion, the reality would be two or three times more than the information reported by the Government. In fact, many international institutions evaluated the accident at Fukushima, and the result was actually two or three times higher. I think these results are the truth. In short, 400 or 500 times the nuclear fallout from the Hiroshima bomb was released into the atmosphere from the Fukushima Daiichi NPP. This is a serious catastrophe.

How did the radioactive fallout flow?

Japan is situated in the temperate region of the northern hemisphere, where the prevailing westerlies blow strongly. Basically, the wind blows from the west to the east in Japan. The Fukushima Daiichi Nuclear Power Plant is located on the pacific shore of the Fukushima prefecture, and its east side is the ocean. Because of that, the massive amount of radioactive substances were transported by westerlies and released into the atmosphere, which flowed into the ocean. Of course nature is not one-dimensional. Also north, south and easterly winds blew at the time of the accident.

According to the Government's report, the radioactive substances blown by the north wind heavily contaminated the area of Hamadori in Fukushima. Then, the radioactive substances transformed into plume and passed over the prefectural border, continuing to pollute the northern part of the Ibaragi prefecture. After that, radioactive plume left towards the Pacific Ocean but came back to the land towards the southern part of the Ibaragi prefecture and polluted the area of Kasumigaura. The radioactive substances polluted the northern part of Chiba and a part of old downtown Tokyo.

There were also times the wind blew from the southeast, and transferred the radioactive substances to the northwest. Unfortunately, it was raining and snowing in the area at that time. Consequently, the radioactive substances, mixed with rain and snow, have fallen and moved forward by severely polluting the land. The path became designated areas such as "the advisory evacuation lift preparatory area", "habitation restriction area", and "difficulty in returning area".

After advancing 40 to 50 km, the wind direction changed. This time the north wind started to blow and the radioactive substances were transported to the south. Later, the radioactive plume swept the streets of Nakadori, polluting the most populated area in Fukushima. Eventually, the plume reached the border with the Tochigi prefecture. It passed over the border easily, since there are no barriers or anything and polluted the northern half of Tochigi. Moreover, it passed over the border of the Gunma prefecture and continued polluting the northern half of Gunma. The wind would have reached Nagano if it had continued straight, but the mountains of high altitude prevented it from advancing. Then the wind turned down to the south along the mountains, engulfing the west of Saitama as well as Okutama of Tokyo.

Let me describe the word "contamination" with specific numbers. The area northwest, 40 to 50 kilometres away from the Fukushima Daiichi NPP exceeds 600,000 Bq/m^2 (Becquerel per square metre). In the Nakadori area, the Government declares 100,000 to 300,000 Bq/m^2 was detected. In other areas: the east half of Fukushima, the northern half of Tochigi and Gunma, the southern and northern part of Ibaragi, the northern part of Chiba, a part of Saitama and Tokyo, are about 30,000 to 60,000 Bq/m^2. Furthermore, in addition to what I have pointed out, some areas such as the southern and northern part of the Miyagi prefecture, a part of Niigata and a part of Iwate were also contaminated about 30,000 to 60,000 Bq/m^2. Let me emphasise that all the data is reported by the Japanese Government.

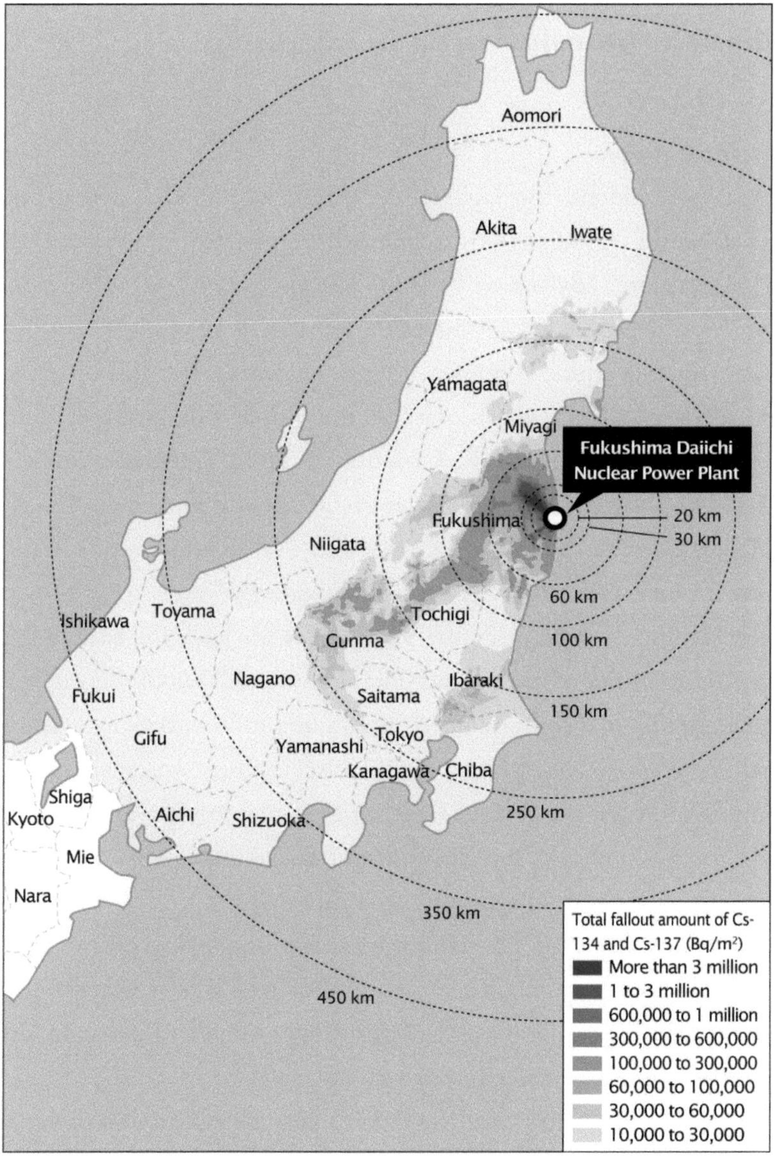

Source: Based on the document from the Ministry of Education and Science

Figure 9: The areas of radioactive contamination

Radiation controlled areas in expansion

Abandoned people

After the accident, the Japanese Government initially advised people who lived within a radius of three kilometres from the Fukushima Daiichi NPP to evacuate. *"Buses will be sent. Take the bus and go to the evacuees' centres."* So the people left as advised.

Later, the Government informed residents within a radius of ten kilometres, *"please evacuate just in case."* So the people took buses and left for the evacuees' centres. Then, the Government said, *"People who live within a radius of twenty kilometres should evacuate just as a precaution."* So those people left by buses. Finally, the Government said, *"people who live within a radius of thirty kilometres may evacuate voluntarily."*

Here we'll look again at Figure 9 (page 53) to check the contaminated areas within a radius of twenty and thirty kilometres. 60,000 Becquerel was detected even in low contaminated areas, and 3 million Becquerel in high-contaminated areas. Normally, people must be forced to evacuate immediately. In addition, the damage from 3.11 did not only include radioactive contamination, but also the gigantic earthquake and the tsunami attack. The towns, cities and villages were in a catastrophic state. Electricity was down, stores were closed, gas stations were closed, and the roads were blocked off. If buses came to pick them up, they would have evacuated. But many people were certainly not in a position to be able to evacuate just by being told to go somewhere if they wanted to.

Logically, radioactive substances didn't spread evenly around the nuclear power plant, but rather they were scattered around by wind transportation. The Japanese Government hid this information despite the report from SPEEDI (System for Prediction of Environment Emergency Dose Information). Instead, they advised people to evacuate by drawing mechanical circular lines by distance from the nuclear plant. The biggest harm caused by this decision was to the habitants of the Iitate village. The Iitate village is located at the edge of the area extending northwest from the Fukushima NPP. It is outside of the thirty-kilometre radius from the site. Due to previous advice from the Government, the people in Iitate were suddenly told to leave from their village several months after the accident. The Iitate village was famous for its beautiful scenery. It was known as an independent village refusing to receive any kind of benefits from the nuclear industry.

What is a radiation controlled area?

I work at the Research Reactor Institute of Kyoto University. There is a nuclear reactor there. The size of the land is 100,000 tsubo, which is about 33 hectares. The nuclear reactor is tiny like a gadget, but it requires such enormous land around it. Furthermore, when we operate the reactor, it should not be operated anywhere within 100,000 tsubo, but has to be done at a designated place. This specific place is called a radiation-controlled area. Access to the area is strictly restricted. The only possible place for ordinary people to be in contact with a radiation-controlled area is at a hospital. X-ray rooms and CT scan rooms are radiation-controlled areas. When you enter such rooms, you might notice the signs saying, "Authorised persons only". Only people with special permission can enter these places.

Once you are in a radiation-controlled area, you are not allowed to drink, eat or sleep. The door of a radiation-controlled area is always closed, and you must follow a special procedure to go through the door.

Working inside the area means that your clothes will be contaminated; your hands will be contaminated. If you leave the area in such conditions, you might irradiate other people outside. Therefore, the system does not allow the door to open unless the person inside decontaminates himself completely. If their clothes are contaminated, they have to be thrown away. If their hands are contaminated, they have to be cleaned. If the water doesn't clean the hands enough, you must use hot water. If that's not enough, you must use chemical detergent even if it peels off your skin. The door won't open unless the body is completely clean.

The level to be able to go through the door is 40,000 Bq/m^2. The bottom line is that no one is allowed to take out radioactive substances exceeding 40,000 Bq/m^2.

Is Japan a constitutional state?

If an area of 40,000 Bq/m^2 is considered a "radiation-controlled area", the severely contaminated areas such as: 40-50km Northwest of the Fukushima NPP, located inside a radius of 30km, should be designated as a radiation-controlled area too. Moreover, the areas contaminated with 30,000 to 60,000 Bq/m^2 should also be considered radiation-controlled areas. In particular: part of the Iwate prefecture; the south and north of the Miyagi prefecture; the east side of Fukushima; the north side of the Tochigi prefecture; the Gunma prefecture; a part of the Niigata prefecture; the north and south of the Ibaragi prefecture; the north part of Chiba; a part of the Saitama prefecture; and Tokyo.

The area of 60,000 Bq/m^2, where the surface is 1.5 times larger than the lake Biwa, has no habitants because the Government already forced people to evacuate. If the rest of the areas are added, it will be more than ten thousand km2 of land. Today, in this enormous radiation-controlled area, several million people live, give birth and raise children.

What can we do? I no longer know. Such a tragedy, an unimaginable tragedy is still going on at this very moment in Japan.

It is illegal to take objects that exceed the contamination of 40,000 Bq/m^2, outside of a radiation-controlled area. So, if the Japanese law is to be respected, the sites, which exceed 40,000 Bq/m^2, should be designated as radiation-controlled areas, and access should be restricted. However, because the Japanese Government cannot abandon the land of some ten thousand km^2, they have decided to leave the people there. Apart from that, there is a law that restricts people from being exposed to more than one mSv (millisievert) of radiation annually. In reality, as the Government reported, not only the Tohoku region but also the vast areas in Kanto are heavily contaminated. Faced with this reality, what did the Government do? They said, *"Before, the situation was normal. Now we tolerate annual radiation of 1 mSv due to the emergency situation"*. Irradiation is a high risk, especially, to children whose cell division is active. To avoid irradiation, the only solution is to abandon the land and leave. But the Government abandoned the people who are in the category of voluntary refugees, without any compensation or support. Some who had resources, left. Others, who could not leave with their entire family, tried at least not to irradiate the children, sending only the children and mothers away while the fathers stayed for work. For farmers and dairy producers, the land is their livelihood. They cannot leave the land behind. Many people, including children, were abandoned within the areas, which should be designated as a "radiation-controlled area".

There is a water concentration regulation when contaminated water is drained into natural water. Further back in the past, for example, at the Rokkasho Village Reprocessing Plant, which is still under construction, this regulation was ignored, as it is too strict to be respected. Instead, the drainage pipes were extended to three kilometres away from the shore, then the polluted water was drained 40 metres below sea level by diluting it with seawater.

Just six months ago or so, in Minami Soma city, the polluted water, which exceeded the regulation level, was drained into the river. The Japan Atomic Energy Agency accepted this fact.

Japan has been calling itself a constitutional state. When citizens break the law, the State punishes them. Then, I think it is at least a minimum requirement that the State respects the laws it has made itself. But the law falls by the wayside when the Japanese Government considers it as too strict to be followed. Considering such problems, I wonder why the laws cannot punish the criminals who caused the Fukushima accident. It is normal to bring those people who are responsible for the accident to justice.

Black particles also found in Tokyo!

In the areas that are supposed to be designated as radiation-controlled areas, there was not only invisible pollution in the air and soil, but also, what are known as "black particles". Research on this phenomenon has been rapidly conducted by Mr Tomoya Yamauchi at the Kobe University. He discovered that the black particles contain extreme concentrations of cesium, and in fact, they are dead cyanobacteria (blue-green bacteria).

Subsequently, the black particles began to be delivered to me around spring 2012. At first, I was asked to measure six samples coming from Higashi Murayama city (two samples) and Katsushika-ward (four samples) in Tokyo.

When we talk about contamination spreading over areas, Tohoku and North Kanto immediately come to our mind. In fact, if we look at the total amount of cesium fallout per town from March to June 2011 (Figure 10), in the data released by the Ministry of Science and Education and the Nuclear Regulation Authority, Tokyo came fourth highest excluding the Miyagi prefecture (where the measurement was not detectable), and the fifth highest if the Miyagi prefecture is included. From April to October 2013 (Table 3), Tokyo came fourth after Fukushima, Ibaragi, and Miyagi prefectures.

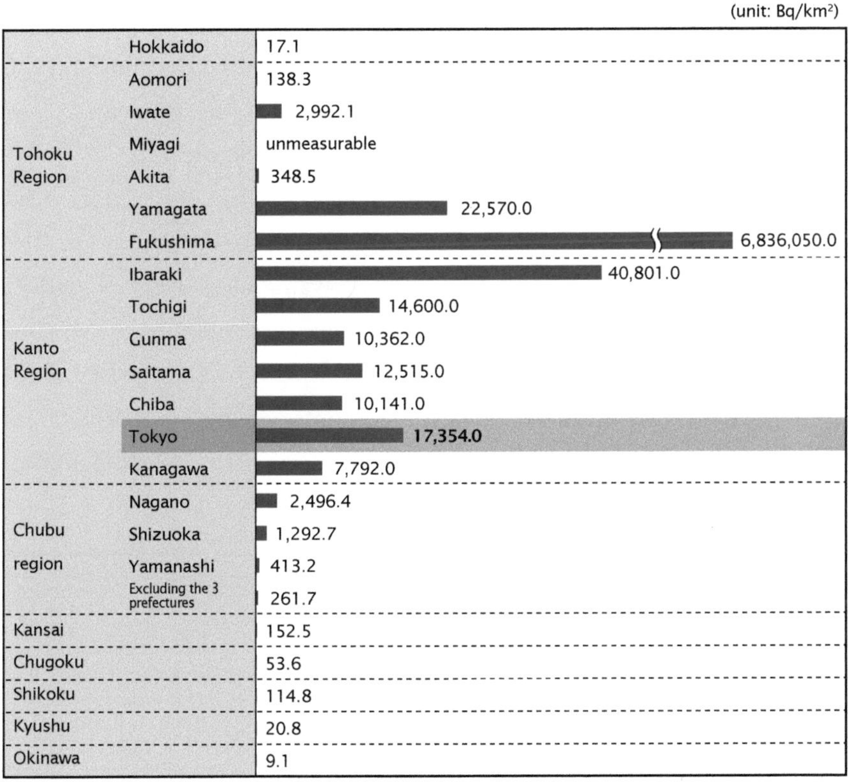

Source: Based on the document "the amount of fallout and contamination of radioactive substances by prefecture between March and June 2011", Ministry of Education and Science (as at December 2011)

Figure 10: Volume of cesium 134 and 137 fallout (March to June 2011)

(unit: Bq/km^2)

1	Fukushima (Futaba)	11,435.00
2	Ibaraki (Hitachinaka)	143.40
3	Miyagi (Sendai)	49.69
4	Tokyo (Shinjuku)	48.75
5	Gunma (Maebashi)	39.73

Source: Based on the report "Result of the investigation on the level of environmental radiation (monthly fallout)" by the Nuclear Regulation Authority

Table 3: Volume of cesium 134 and 137 fallout (April to October 2013)

Table 4 is the list of samples taken including the extracts from Tokyo. Table 5 indicates the concentration of cesium. As I wanted to find out the concentration level of cesium, the measurement was taken by a gamma ray spectrometry by using a germanium semiconductor detector.

The black particles found in the Mizumoto Park in Katsushika-ward where the old towns are located, contains 20,000-30,000 Bq/kg. The cesium found at the school in Higashi-Murayama city (Tokyo 1 and 2) contains over 20,000 Bq/kg; and the samples extracted from Minami Soma city, around the same time, contained 2 to 6 million Bq/kg of cesium. Furthermore, in summer 2013, the samples arrived from other areas in Fukushima outside of the mandatory evacuation area. Nearly 2 million Bq/kg cesium was detected in those samples.

Sample	Sample site	Measured date	Weight of the sample (g)	Measured duration (min)
Tokyo 1	Higashimurayama, school	30/5/2012	54	76.2
Tokyo 2	Higashimurayama, school	30/5/2012	57	100.7
Tokyo 3	Katsushika, Mizumoto park	30/5/2012	61	8.1
Tokyo 4	Katsushika, Mizumoto park	30/5/2012	51	11.3
Tokyo 5	Katsushika, Mizumoto park	30/5/2012	72	9.55
Tokyo 6	Katsushika, Mizumoto park	30/5/2012	65	9.93
Fukushima, Minami-soma 1	Kawabusa-oku	7/6/2012	0.68	6.28
Fukushima, Minami-soma 2	Kanaya-shoten-mae	7/6/2012	1	9.8
Fukushima, Minami-soma 3	Kanaya-shoten-mae	7/6/2012	1	16.7
Fukushima, Iitate 1	New Track Iitate	24/6/2013	41.5	1
Fukushima, Iitate 2	New Track Iitate	10/7/2013	0.71	5
Fukushima, Iitate 3	New Track Iitate	10/7/2013	0.86	60
Fukushima, Iitate 4	Nagadoro	10/7/2013	0.88	5
Fukushima, Tamura 1	Takashibayama	10/7/2013	1.29	31.8
Fukushima, Tamura 2	Takashibayama	10/7/2013	1.35	30
Fukushima, Fukushima	Sanbongi, Healthy Land	10/7/2013	1.3	30

Table 4: List of samples "black particles"

Now let's look at how much those areas from where the samples were extracted were polluted. Mizumoto Park in Katsushika-ward should be designated a radiation-controlled area because it is contaminated by more than 40,000 Bq/m^2. Minami Soma city and Fukushima city are contaminated areas exceeding 100,000 Bq/m^2. Meanwhile, Higashi Murayama city is not polluted enough to be classed as a radiation-controlled area, however, the black particles were found in the playground at the school.

(en Bq/kg)

Sample	Cs-134	Cs-137	Total
Tokyo 1	8,500 ± 59	14,000 ± 80	22,500 ± 99
Tokyo 2	8,400 ± 51	13,000 ± 66	21,400 ± 83
Tokyo 3	110,000 ± 720	170,000 ± 870	280,000 ± 1 130
Tokyo 4	110,000 ± 640	180,000 ± 860	290,000 ± 107
Tokyo 5	89,000 ± 500	150,000 ± 680	239,000 ± 844
Tokyo 6	74,000 ± 470	120,000 ± 590	194,000 ± 754
Fukushima, Minami-soma 1	190,000 ± 14,000	3,800,000 ± 21,000	5,700,000 ± 25,200
Fukushima, Minami-soma 2	760,000 ± 6,100	1,500,000 ± 8,800	2,260,000 ± 10,700
Fukushima, Minami-soma 3	990,000 ± 5,400	1,900,000 ± 7,600	2,890,000 ± 9,320
Fukushima, Iitate 1	2,400,000 ± 32,000	4,760,000 ± 49,000	7,160,000 ± 58,500
Fukushima, Iitate 2	4,260,000 ± 25,500	10,400,000 ± 40,560	14,660,000 ± 47,900
Fukushima, Iitate 3	2,020,000 ± 4,650	4,980,000 ± 7,470	7,000,000 ± 8,800
Fukushima, Iitate 4	2,870,000 ± 20,400	6,790,000 ± 29,200	9,660,000 ± 35,600
Fukushima, Tamura 1	129,000 ± 1,300	318,000 ± 2,000	447,000 ± 2,380
Fukushima, Tamura 2	403,000 ± 2,340	996,000 ± 3,680	1,399,000 ± 4,360
Fukushima, Fukushima	563,000 ± 3,000	1,340,000 ± 4,400	1,903,000 ± 5,330

Table 5: Concentration of cesium in the samples of "black particles"

Under Japanese law, every object, which exceeds a cesium 134 and 137 contamination of 10,000 Bq/kg, is considered as radioactive material. Therefore, normally, these materials must be treated strictly inside a radiation-controlled area. In reality, they remain in the areas where people live. The black particles must immediately be removed from the places where children play. The Government and the authorities must at least deal with this issue by following the laws they established.

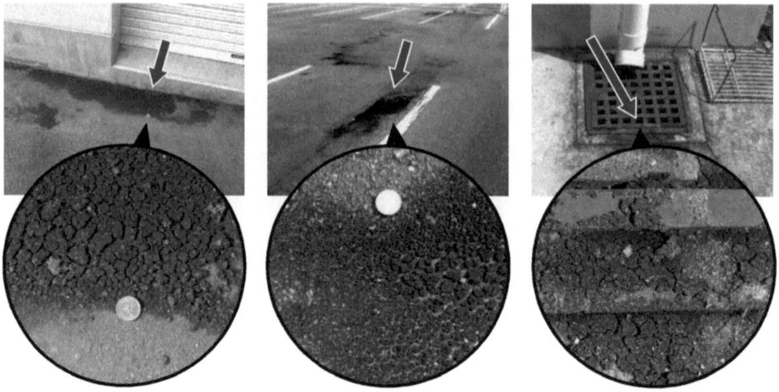

Temple Butsugen, Iwasaki-cho, Watari, Fukushima (photo taken on 21 April 2012)

New Track Iitate (photo taken on 30 April 2012)

Nagata, Hirata, Fukushima (photo taken on 28 March 2012)

Photographs: Shigeki Chiba, Mr. Shigeki Chiba, teacher at Ono High School of Hirata, Fukushima extracted from "The report on mass spectrometry measurements by accelerometer, Nagoya University" (XXIV), March 2013)

Photograph: "Black particles" in our everyday environment

People who don't give up

Decontamination project only causing sacrifices to the population

With the hope that the vast contaminated land could somehow be recovered, the State has invested an enormous amount of national expense for the decontamination project. The reconstruction budget (the entire list of the project budget for the Great East Japan Earthquake reconstruction for financial year 2013) shows the amounts of 454.7 billion yen in financial year 2012 and 622 billion yen in financial year 2013, as you can see in the column titled "decontamination and others". The total cost will be approximately one trillion, one hundred billion yen. Each year, a little less than half of this budget has been accumulated in the "fund" managed by the Fukushima prefecture, in order to spend it on decontamination for the areas where the State does not directly get involved. In financial year

2012, less than 40% was spent from the fund only, which means that the decontamination did not progress a lot. Consequently, the remaining 60% is automatically transferred to the following year. Curiously, almost the same amount of new budget was added on top of the previous budget.

Why doesn't the decontamination project advance to begin with? This question was put to the person responsible for the local municipality in an article posted in the Asahi Digital version (12 July 2013). According to the article, it was difficult to proceed with the decontamination due to the strict guidelines imposed by the Ministry of the Environment. At the same time, the article points out *"the decontamination will be accelerated if the decision-making is handed over to municipalities."* Affirming this with the Ministry of the Environment, they replied, *"Techniques that are not approved by TEPCO cannot be accepted."* TEPCO covers the decontamination cost which the State doesn't control directly. The Ministry of the Environment has made a claim for TEPCO to repay the decontamination cost of 40.4 billion yen that was paid in advance, but TEPCO has not paid it back yet.

To the Ministry of the Environment, it is logical *"to negotiate to avoid the situation that TEPCO refuses to pay"*. This is absurd.

To summarise, the State is more concerned about saving TEPCO than saving its own people.

In December 2013, the Government decided to cover all expenses except for decontamination, as well as the equipment and construction costs for the temporary storage installations for decontaminated waste. The decontamination project may not finish for several decades or even one hundred years. The expected cost is estimated at five trillion yen, but there is no certainty. No matter how much cost there will be, the State will be in charge of payment. Just before this decision was made, the estimated five trillion yen was decreased by half (2.5 trillion yen). How did this happen?

Some of you may recall this breaking news. In November 2013, the previous month, the Nuclear Energy Safety Organization announced that irradiation below "20 mSv per year" has no strong impact on health. In fact, this report is related to their later decision.

Here is the summary of how it works: TEPCO needs to be reconstructed → the State will cover the decontamination costs → but it costs too much → how can we make it cheaper → maybe the size of decontamination operations can be reduced? → I know! Instead of attempting to decontaminate 1mSv/year, let's change the rule and accept irradiation of up to 20 mSv/year. As a consequence, it is the lives of people who are sacrificed. The Government also proposed a "support plan" to give the residents who evacuated from the "difficult to return areas" the choice of returning home. This is also a strategy to save the decontamination costs. In areas which are highly contaminated it takes time and money for decontamination, therefore, they have decided to leave these areas as they are, and concentrate on the less contaminated areas.

There is a sequel to the story: the cost will now be reduced to half, but it still costs 2.5 trillion yen → I know! Let's use the profit from the sales of TEPCO shares held by the Minister of State for the Corporation in Support of Compensation for Nuclear Damage. The "profit from sales of shares" is the profits gained when the stocks are sold in the future. Today, the value of TEPCO shares is valued at one trillion yen. The Government estimates the value will increase to 2.5 trillion yen (2.5 times more) in the future. But the future of TEPCO is highly uncertain. If the value does not go up as the Government expects, the loss has to be compensated by our tax or electricity charges. Also, as the estimate is planned without an actual profit, it requires the financial resources to pay the costs now. Where do these financial resources come from? It is known as the "promotion for development of power resources tax", which we pay with our electricity charges. In any case, it is people like us who pay.

TEPCO's arrogance

TEPCO, with a guarantee by the Government who offered to pay the decontamination costs, now has a prospect. They stated that they would reduce the electricity costs in ten years if the Kashiwazaki Kariwa Nuclear Plant restarted in July 2014 as planned.[6] It means that they are still claiming that nuclear energy is cheaper than other energy resources. I will explain in detail about the electricity costs in Chapter Five. In my opinion, this view is totally out of the question! Lowering the costs is not because nuclear energy is cheap, but is simply related to the issues within the company's management.

On 16 January 2014, Naomi Hirose, the president of TEPCO, went to see Mr Hirohiko Izumida, the Governor of Niigata as a New Year's greeting. This meeting was held in public audience. At this meeting, the Governor Izumida criticised TEPCO's reconstruction plan as it forces consumers to choose between *"restarting [the nuclear power plant] or raising electricity costs"*. He pointed out that the plan is a *"moral hazard that takes the responsibility away from stockholders and lenders (financial institutions)"*. This manner of escaping from the responsibilities, which fall on those involved in the decision-making, indicates, *"Despite the accident, they don't have to accept any responsibility"*. He expressed his anxiety as such attitude may promote higher risks when operating a nuclear plant. The Governor also criticised TEPCO's difficult financial situation as it was caused by the Fukushima accident. Without a thorough examination of the accident or a construction of *"the governance system to prevent an accident"*, it is impossible to reconstruct a nuclear plant, he commented.

There was more news: TEPCO announced that they would separate the decontamination projects and create an independent company in order to specialise in the project from April 2014. This announcement was probably

[6] This arbitrarily fixed calendar was obviously not respected. As at the end of November 2014, the plant has not restarted and the possibility remains uncertain.

to show the Nuclear Regulation Authority, which functions as a judging mechanism for restarting a nuclear plant, that *"we can concentrate on the operation of the nuclear plant"*, and *"we have nothing to do with the after treatment of the accident, because the other company is in charge of it"*. It is such an arrogant attitude, I can hardly bear to listen to it.

On the other hand, TEPCO was trying to promote a good image to the victims in Fukushima. They decided to relocate some of the members in management positions who were over fifty years old to Fukushima. (Sankei Shinbun, 21 December 2012). They were dismissed from their positions in exchange for a pay raise. On the contrary to the young employees who quit TEPCO in despair, those dynamic older employees have experience and motivations. By sending them to Fukushima, TEPCO attempted to encourage the workers at the site. Actually, those people will probably not be sent to the risky sites and of course, the board members are not included as candidates.

This is how it works. The people considered to be the "top dogs" at TEPCO, the Japanese Government, and bureaucrats, remain "clean". Then all the hard work or painful charges are pushed onto the people in weaker positions or people who are not able to challenge authority. This culture has been repeated time after time. Even during the crisis caused by the Fukushima NPP accident, they never stopped escaping from their responsibilities. We have come to the point when this ugly habit must to be stopped. Because of this reason too, we should not accept that the nuclear power plant in Japan be restarted.

The "new standard", just a procedure in order to restart nuclear plants

Restarting a nuclear plant first of all requires respecting the new regulatory standard (new standard) proposed by the Nuclear Regulation Authority. The new standard was implemented in July 2013. Ever since, sixteen units among nine nuclear plants including the Kashiwazaki Kariwa

Nuclear Power Plant owned by TEPCO have been registered for compliance inspection. (Table 6)[7]. Ten units among the six nuclear plants - Tomari, Oi, Takahama, Ikata, Sendai and Genkai – have been undergoing inspection. As a result of the inspection, Kashiwa No.3 and Takahama No.3 and 4 are now under construction to meet the standard.

Electric Power Company	Subject reactor	Applied date	Progress status
Hokkaido Electric Power Company	Tomari, reactors No.1 and 2	8 July 2013	Inspection on hold
	Tomari, Reactor No. 3	8 July 2013	Approval expected (renovation of cooling device in progress)
Kansai Electric Power Company	Oi, Reactors No. 3 and 4	8 July 2013	Approval expected
	Takahama, Reactors No. 3 and 4	8 July 2013	Approval expected (construction of tide embankment in progress)
Shikoku Electric Power Company	Ikata, Reactor No.3	8 July 2013	Approval expected
Kyushu Electric Power Company	Sendai, Reactor No 1 and 2	8 July 2013	Approval expected
	Genkai, Reactor 3 and 4	12 July 2013	Approval expected
Tokyo Electric Power Company	Kashiwazaki-Kariwa, Reactors No. 6 and 7	27 September 2013	Inspection on hold
Chugoku Electric Power Company	Shimane, Reactor No. 2	25 September 2013	Inspection in progress
Tohoku Electric Power Company	Onagawa, Reactor No.2	27 September 2013	Inspection in progress

As at the end of January 2014

Source: Based on the "inspections of compliance with the new regulatory standards (nuclear power plants)", HP of the Nuclear Regulatory Authority (the progress status was taken from the article Sankei Digital on 20 January 2014 and Asahi Shimbun on 5 February 2014)

Table 6: The nuclear power plants that have applied for an inspection on compliance with the new regulatory standards

[7] Among the power plants registered, the Sendai Nuclear Power Plant (reactor No.1 and 2) owned by the Kyushu Electric Power Company has been approved to be restarted.

In January 2014, the Nuclear Regulation Authority announced, *"The first approval will be given as soon as this spring"*. (Sankei Digital, 20 January 2014).

The Nuclear Regulation Authority proudly claims this new standard is *"the strictest standard in the world"*. I don't understand how they could come up with this new standard. The reason they had to make a new standard was because of the accident at the Fukushima Daiichi NPP. At the site, no humans have had access to the reactor building yet, due to the massive amount of radiation, thus no one knows the extent of the damage in there, not even where the nuclear cores are. No one knows the real cause of the destruction of the reactors; whether it was due to the earthquake, the tsunami, or something else. How can a new standard be created without first finding out the answers to these questions? My estimate for the investigation on the accident would take at least ten years. In the meantime, it is not possible to create a new standard, not to mention to restart a nuclear plant.

The new standard gives us the impression that there will be a secured nuclear plant. By calling it not just a "new safety standard" but also a "new regulatory standard" already indicates that the nuclear plant is not guaranteed to be absolutely safe, as proven by the Fukushima accident. This standard does not mean, "the operation will be safe if it is respected", but rather "restarting will be accepted if it is respected." Realistically, it is impossible to come up with a plan, as they don't even know the cause of the accident to begin with. Besides, those who have made the new standard and manage it are members of the so-called "Nuclear Mafias". The Nuclear Regulatory Authority is an institution, which evolved by separating from the Nuclear and Industrial Safety Agency (NISA) in September 2012. For example, at the beginning of February 2013, one of the administrative members in the Nuclear Regulatory Authority secretly passed an internal document, a draft report for evaluating earth faults, to the Japan Atomic Power Company (JAPC). I wasn't shocked at all. They just did what they are used to doing. Fundamentally, nothing has changed.

A nuclear plant is a gigantic machine. There are multiple possibilities a machine could fail. Mankind has already experienced the accidents of melting reactor cores and the release of massive amounts of radioactive substances, four times in our history. The first accident happened to a Windscale plutonium generating reactor in England in 1957. The second accident took place at the nuclear plant in Three Mile Island, USA. Then there was the accident in Chernobyl in the USSR in 1986. The latest one is the accident at the Fukushima Daiichi NPP. The Fukushima accident was triggered by the earthquake and tsunami, but the other three had nothing to do with a natural disaster. Yet, the causes of all four events exceeded our estimations. If a fifth event occurs, it will be due to another cause. This is the nature of the machine. For example, the Nuclear Regulatory Authority, when examining a nuclear plant, might advise that unit 3 and 4 of the Takahama Nuclear Power Plant re-consider the level of a potential tsunami attack to be higher than before. Kansai Electric Power Company is rushing into the construction of a tide embankment. But even if they would raise the estimation of a tsunami to the maximum, it is ridiculous to conclude that "safety is guaranteed". Also, the New Standard allows certain "delay" for facilities such as a secondary control room, which takes a long time to construct. However, the accident is unpredictable. It may not require the function of a secondary control room. Or it may be absolutely necessary.

An accident occurs outside of our estimations. A nuclear accident, once it occurs, produces enormous damage, as we have witnessed at Fukushima. In principle, I always believe that we should abandon the nuclear industry completely. However, if measures are to be taken to control it, these measures should at least restrict a restart unless a candidate fulfils all requirements. If the Nuclear Regulatory Authority cannot even stand by this, the new standard will be discounted as "just a procedure for restarting nuclear plants".

"Reconstruction": a gift for major construction companies

At the end of September 2013, an international exhibition called "decontamination of radioactivity in the environment/contaminated waste treatment" was held in Tokyo. At this exhibition, enterprises and research institutions from Japan and overseas, which deal with the decontamination and/or radioactive waste treatment industries, were gathered and presented their latest technologies and products. The size of the exhibition was quite significant as more than one hundred enterprises participated during the three days.

The following were on the list of exhibitors (in alphabetical order): Areva from France at the top, followed by constructors such as Kashima, Shimizu, Taisei, Toshiba, Japan Atomic Energy Agency (JAEA), the Japan Atomic Power Company (JAPC), Hitachi, Maeda, Mitsubishi-materials... Every one of them was a former participant for promoting the nuclear industry in Japan. In addition, the financial source for the decontamination operation is, as previously stated, the promotion of electrical development tax. This tax was originally established to promote nuclear energy. Those who made profit by constructing nuclear plants are now making profit with decontamination businesses. The lobby of promoting the nuclear industry maintains the same position.

On the other hand, the areas where the Government does not directly operate in decontamination, municipal officers are following the Government's policy, and giving priority to local companies for decontamination businesses. As a consequence, the construction companies in charge of the decontamination businesses profit from their "decontamination special order". The Fukushima Constructors Association and seven other branches have donated 4.88 million yen to the political fund organization of LDP (Shukan-Asahi, 13 December 2013). Objectively speaking, it was an expression of gratitude.

Decontamination is not the only business. In the disaster areas, there are numerous projects that benefit from the national budget such as constructions of public installations, collective residences and infrastructures. "Reconstruction" became an unexpected gift to major constructors as well as to local entrepreneurs. Once they profit from "the gift", they want to return something to ensure more gifts to come. Within the budget of financial year 2014 for reconstruction, five hundred billion yen was given to decontamination businesses. One trillion, seven billion yen was used as gifts called "urban renovation and reconstruction", which make up a total of two trillion, two billion yen. Needless to say, competition for concessions to win these juicy gifts continues even today.

Decontamination is only a waste of national budget

Corner-cutting decontamination spreads pollution

When prime contractors such as major constructors and local construction companies receive an order, they pass the project onto sub-contractors by promising them a share of the profit. Responsibility is also passed onto the sub-contractors. From that point, the primary contractors will never check or look at the sites. The sub-contractors use lower sub-contractors when they are overloaded with projects, and/or when there are projects that they don't want to deal with. This flow of passing down projects and responsibility stems from primary contractors to sub-contractors to sub-sub-contractors and so on and so on. The profit each contractor can receive is reduced as they go further down the line, while the work becomes harder and harder. In addition, their work is fragmented into pieces. The position of responsibility becomes unclear. Nonetheless, the work must be finished within fixed periods, and if they finish earlier than expected, they will receive more benefits. The pursuit of efficiency with the minimum amount of labour results in corner-cutting decontamination.

For example, the water used for cleaning roofs of individual houses by high-pressure hoses must be stored in one place instead of discharged onto the ground. If this rule is respected, the decontamination would consume too much time and money, therefore, the water is discharged onto the ground. There were other kinds of reports such as: contaminated deposits being physically kicked off (from branches etc.) into the rivers; used tools being washed in the rivers; dirty mud being dumped into gutters after boots were washed, etc.

In December 2013, the workers dug a huge hole in the backyard of the house that was "decontaminated". They threw used cloths after wiping down radioactive substances, wires, plastic bags, cables, glasses, metal poles and wooden objects and even bicycles into the hole (Shukan Asahi, 13 December 2014).

According to a report from the workers at the sites, even a "secret manual for corner-cutting" existed in order to deceive inspectors' eyes that only patrol around certain hours and places. Furthermore, it appears that there was somewhat of a familiarity and cosy relationship between workers and inspectors (Shukan-Asahi, 20 December 2013). Around January 2013, when corner-cutting decontamination was being revealed successively, the Ministry of the Environment announced that inadequate sub-contractors would be banned from bidding. According to this scoop article, none of the sub-contractors were denounced or banned from bidding after all.

The aim of decontamination is to gather and isolate radioactive substances from residential areas. However, in reality what is happening is they are not being isolated, but rather they are being spread. It's not simply meaningless, but it's causing more harm than good. Without a doubt, the corner-cutting decontamination continues even today.

No need for decontamination up to "20 millisievert/year"

"Decontamination" means getting rid of contamination. In reality, it is impossible to eliminate contamination. What can be done is to transfer the contamination to elsewhere. Thus, I use a term "trans-contamination". Apart from artificial trans-contamination by humans, nature also transfers contamination. Contamination in the mountains slides down to towns, contamination in the rivers flows into the ocean. Environments are naturally linked to each other. Therefore it is impossible, even without corner-cutting decontamination, to stop the radioactive substances from spreading to non-contaminated places. The areas that have already been contaminated would be re-contaminated again. Blocking radioactive contamination is fundamentally impossible. In fact, there was a case in Matsudo-city (Chiba prefecture), where the amount of radioactivity in the park, which was once decontaminated, exceeded the standard level again.

The Japanese Government propagates that decontamination will allow people to go back to the contaminated areas, and that they have been investing enormous national expenses for decontamination projects. If they are serious, the total areas of contamination should be decontaminated simultaneously, but such an idea is not realistic. The current operational method will only have a limited impact; therefore it's better to be sceptical about the decontamination fantasy.

To begin with, it is not necessary to spend money in order to begin a decontamination project if "20 millisievert (mSv) per year" is the target value. At the moment, cesium is the primary substance polluting lands. There are two types of cesium: 134 and 137. Figure 11 shows the evolution of the atmospheric dose of gamma ray considering the physical half-life period of cesium. At the beginning, cesium 134 (Cs-134) occupies 70% of the space. As its half-life is as short as two years, cesium 134 decreases progressively, and almost disappears within ten years. That means that the total contamination by cesium will decrease by 70% in ten years. If we examine interim progress: half of the 70% will disappear in two years; and the remaining amount will be further halved in four years. Four years have passed since the accident. Today, contamination by cesium 134 will be one fourth of the original amount. In the meantime, cesium 137 also decreases to a small degree. As a result, the estimated amount of cesium in four years will be less than half of the original amount. This is only a calculation of the atmospheric dose, considering external irradiation only.

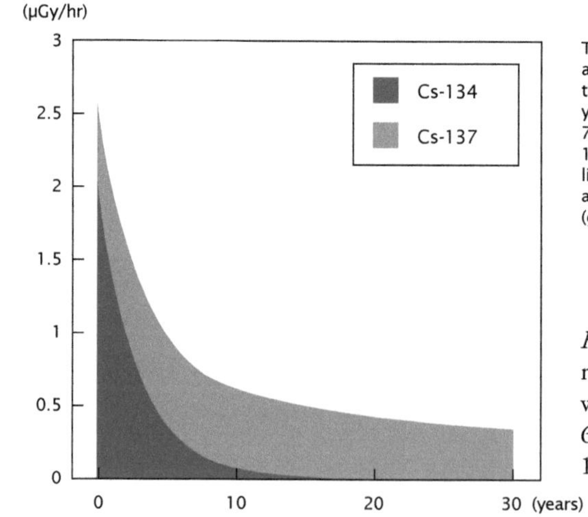

The gamma rays emitted by the Cs-134 are about three times more powerful than those of Cs-137. The half-life of Cs-134 is 2 years and that of Cs-137 is 30 years. About 70% of initial irradiation is caused by Cs-134. However, because of the short half-life of Cs-134, the irradiation remaining after 10 years is mainly caused by Cs-137 (gamma rays).

Figure 11: Evolution of the atmospheric dose of gamma ray within the contaminated areas at 600,000 Bq/ m^2 as a sum of Cs-134 and Cs-137

The Government is planning to manually decontaminate the areas under 50 mSv by spending 2.5 trillion yen, and try to send the people back home by reducing the contamination level to less than 20 mSv/year. But the contamination will be reduced naturally, without any intervention by humans, to about 20 mSv in 2015 compared to 50 mSv/year at the time of the accident; even if the level is at 50 mSv today, in some areas, this will be reduced to around 20 mSv within four years. In six years, it will go down to 20 mSv; and in 10 years to 15 mSv.

Limited dose of internal irradiation for pregnant women: 1 mSv even for workers involved with radioactive operations

Now, let us examine the number "20 mSv/year". Even someone like me, who is paid to work in a radioactive environment, is legally prevented from being irradiated higher than this dose. Researchers have reached the conclusion that irradiation is harmful to the human body regardless of its level. In principle, it is better to avoid irradiation even if it's 1 mSv/year. The radiation workers are expected to tolerate is 20 mSv/year in exchange for their salaries. The Government is telling people, who are not being paid and who are there 24 hours a day, to continue with their ordinary lives because "it's an emergency situation".

The irradiation dose limit for radiation workers is stated in the law called "Ordinance on the Prevention of Ionizing Radiation Hazards", including the following regulation.

> *Article 6. The employer shall take the necessary measures to ensure that pregnant female radiation workers shall not be exposed to a dose that exceeds the value set in each of the following items for the period from the diagnosis of pregnancy to childbirth (hereinafter referred to as «during pregnancy»), as prescribed in the following dose classifications.*
>
> - *1 mSv for the effective dose due to internal exposure.*
> - *2 mSv for the equivalent dose exposure to the surface of the abdomen.*

Even if it is their profession, pregnant women must not be internally exposed to more than 1 mSv/year. Partial exposure to the surface of the abdomen is also limited to less than 2 mSv. If 1mSv is prohibited for pregnant radiation workers, then ordinary women and children who live in the affected area should have no reason to tolerate a higher dose of irradiation.

As long as there are people abandoned in the contaminated areas, it is obviously necessary to take measures in order to reduce irradiation. However, I am against decontamination projects in the first place. Needless to point out its low cost effectiveness, I am fundamentally against the idea that the contaminated areas can be decontaminated. The priority at the moment is to evacuate and migrate people to other places. Leaving their homeland where they have grown up and had lives must be such an agony for the people in the contaminated areas. Nevertheless, the irradiation limit of 20 mSv/year should not be tolerated. Aiming at less than 1mSv/year would take several hundred years. Their homelands are already lost. The Japanese Government is responsible for pronouncing this cruel reality to the people. They should immediately stop campaigning such a fantasy that decontamination will be successful if the State and administration work hard enough, so the people can go home soon. It gives hope to people. The state should propose a policy to transfer entire communities together, instead of letting people move individual bases. Instead of investing enormous amounts of national expense in decontamination businesses, which empowers lobbyists, they should protect and help re-establish the lives of individual taxpayers. I believe that is the most supportive plan for the habitants.

The endless road of nuclear decommissioning

Working on thin ice

How much radiation has sunk to the bottom of the pool?

On 18 November 2013, the task of removing fuel from the spent fuel storage pool at the No.4 nuclear reactor of Fukushima Daiichi began. When the massive earthquake and tsunami hit Fukushima, the No.4 reactor was not active due to routine maintenance. All fuel in the core had been removed to the spent fuel storage pool located in the upper level of the same reactor building. As a result, a meltdown did not happen. However, a significant amount of fuel was left in the bottom of the pool. Specifically, 1,331 spent fuel assemblies and 202 unused fuel assemblies; meaning a total of 1,533 fuel assemblies were left there. If we look at the other reactors that were active at the time of the accident, there were 392 fuel assemblies (spent and unused) in the first, 615 in the second and 566 in the third reactor. The "assembly" is a way of counting assembled fuel rods. In each unit, eight multiplied by eight fuel rods (about 1cm in diameter and 4.2 m long) are stocked. (Figure 12, page 78)

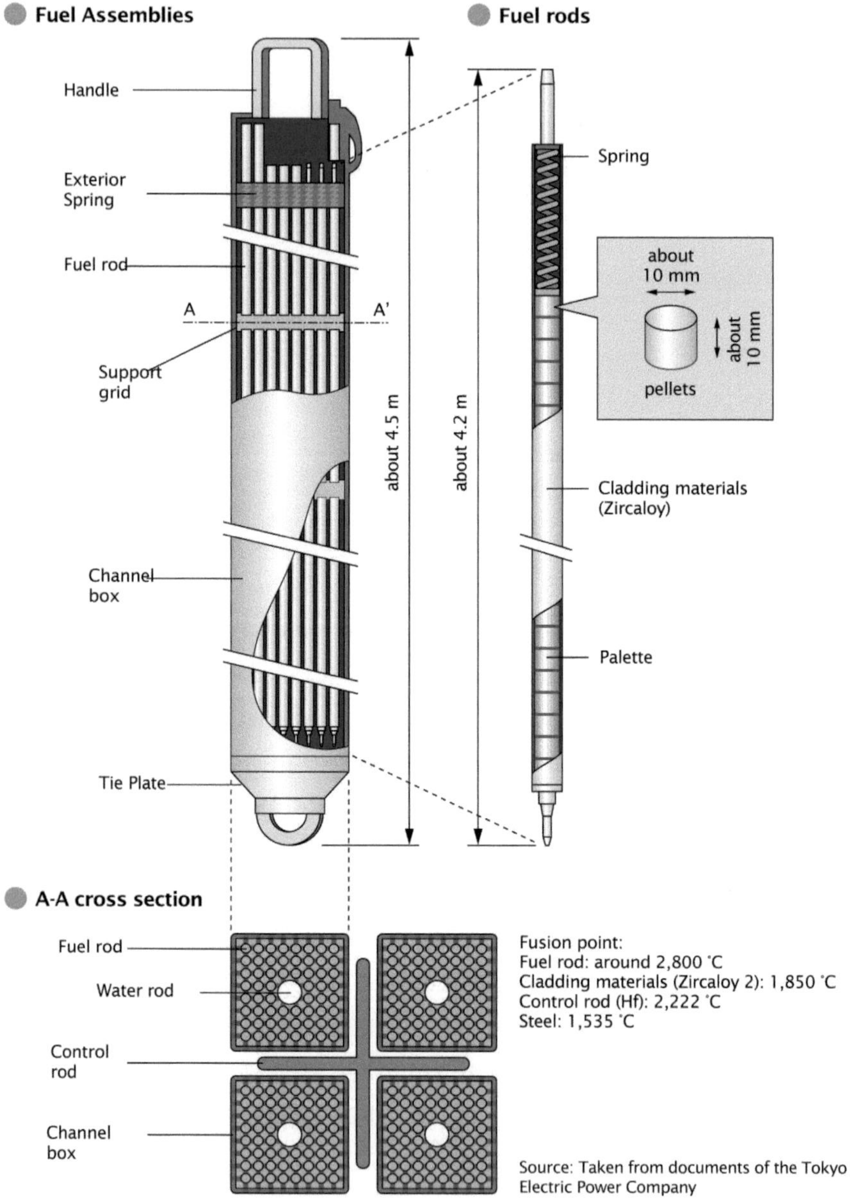

Figure 12: Fuel rods, cladding materials, fuel assembly

Now, let's look at the quantity of radiation in the spent fuel storage pool at the No.4 reactor. Spent fuel means that fuel has reached the limit where it can no longer be consumed. In other words, it is filled with nuclear fission product. It is so dangerous that, if it is pulled out from the water, it can instantly kill people around the pool. There were 1,331 assemblies of them. There was also unused fuel, but the level of radiation was low. One could touch them and not feel any acute symptom of irradiation. In July 2012, TEPCO removed two fuel assemblies in order to examine the extent of the damage. This experiment was only possible because they pulled out the unused fuel assemblies.

In short, spent fuel is the one that threatens us. The estimated total radiation by my calculation, based on the level of cesium 137, is equivalent to 14,000 atomic bombs dropped in Hiroshima. In the previous chapter, I analysed that the diffusion of cesium 137 by the accident, reported by the Government, was 160 times more than a Hiroshima atomic bomb. That means that the quantity equivalent to one hundred times of radiation remains in the bottom of the pool of the No. 4 reactor.

An urgent task: removal of spent fuel from the No.4 reactor

There is another concern regarding the spent fuel storage pool at the No.4 reactor building. The pool is, so to speak, suspended in the air. At the time of the accident, no fuel was left in the core of the No.4 reactor, but there was still an explosion, which damaged a large part of the building. The destruction occurred in a rather curious way. At the No.1 and No.3 reactors, only the operational floors, which were located on the top level, were blown away. But at the No.4 reactor, not only the top floor but also the walls on the floor below were blown away. The spent fuel storage pool was located on that floor.

Immediately after the accident, the media broadcasted the scene of fire engines and the *Self-Defense Forces* dumping a massive amount of water on the building; it was a reaction to control the situation of the spent fuel storage pool. If there was a crack on the surfaces of the pool, which could lead to a leakage of nuclear fuel, the situation could have been a total disaster. Fortunately, the pool itself was not so damaged. However, it was so unstably placed that it would risk a serious accident if there had been another earthquake.

Anticipating the serious situation, TEPCO immediately carried out an emergency operation to reinforce the pool with earthquake resistant supports. As it was a rushed operation under such critical conditions in an irradiating environment, it is doubtful the operation was carried out correctly. Besides, the reinforcement was done on only half of the pool. The other half of the pool is still suspended in the air. Even today, secondary earthquakes continue to hit Fukushima. If the spent fuel storage pool falls down, an unimaginable quantity of radiation will pour out. Concerning these risks, removing the spent fuel that sank to the bottom of the pool to a safer place must be done as soon as possible.

An unimaginably difficult task

According to the work schedule set out by TEPCO, removal of nuclear fuel in the No.4 reactor would be finished by the end of 2014. I believe this project must be completed and should be done as soon as possible. TEPCO is also expecting to carry out this project smoothly and complete it within a year as forecasted. Unfortunately, I don't think it will work out as planned.

Removing nuclear fuel requires the use of a machine called a fuel-handling machine. The spent fuel to be removed is a lump of intensive radioactive substances. Thus, it is impossible to suspend them in the air. It has to be done in the water. First, a huge steel-shipping container called a cask (approximately 5.6m in length, 2.1m in diameter, weighing approximately

91 tonnes), which can stock 23 fuel assemblies, will be submerged to the bottom of the pool. After storing 22 assemblies into the cask, it will be covered and lifted up full of water by a crane, that's how it works, and this process takes about two days. Then the suspended cask, after being decontaminated, will be transferred to a shared pool, which is located 100m away by a trailer. In that shared pool, the reverse procedure will be carried out in order to take out the nuclear fuel. This process will take several more days. To summarise, the entire transfer process takes about a week. TEPCO estimates the whole project will be finished within a year if this process can be done every day using two casks. I strongly hope for their success, but I am somewhat anxious that it will not run as smoothly as planned. In the process of transferring fuel assemblies into the cask, the fuel assemblies may fall or they could hit something and crack. The fuel assemblies might be damaged by the debris, which has fallen to the bottom of the pool due to the explosion. Lifting up 1,533 assemblies without stopping, from a half broken unstable pool is an experience mankind has never encountered in history. There is no guarantee because there is not a single example that ensures its safety. If they fall, the used fuel assemblies will be damaged and radioactivity will jet out. The workers will be irradiated and the work itself may cease. Transportation of the casks is even less stable. The cask alone weighs 91 tonnes. Inserting 22 fuel assemblies weighs nearly 100 tonnes. If it falls during the transportation by crane, the explosion of radioactivity will be massive. The strength of the suspension wires is also a concern. TEPCO announced that the wires are doubled, but if one wire is torn off, will another wire be able to support the cask?

On 15 December 2013, just before the removal of fuel assemblies began, TEPCO announced that there were three broken fuel assemblies in the pool of the No.4 reactor. One of them was bent in the form of a "V", and the other two were leaking. They explained that: the bent one was already being stored in an ad hoc container; and the two leaking ones would be transported to a reliable container which prevents the radioactivity from leaking. Actually, these fuel assemblies were already discovered in 1982, 2006 and 2008. Revealing this fact at this timing is typical TEPCO style. In fact, this damage was before the accident. On top of that, the debris due to the explosion might have damaged more fuel assemblies. Furthermore, even if most debris is removed, the small pieces might damage the fuel assemblies when they are pulled out. If the fuel assemblies are bent or deformed, they won't come out from the container in the first place. Even if they are successfully pulled out, they won't fit into the casks. It is better to imagine other kinds of risks out of our estimation. In any case, it will be an incredibly challenging project, much like walking on ice to pull out each fuel assembly intact, and to complete the total 1,533 assemblies without failing.

Once the removal from the No.4 reactor is finished[8], the No.1, 2 and 3 reactors must also be treated. However, because the No.1 and No.3 reactors have melted down, the radioactive level around the spent nuclear fuel pool is too high to carry out the same project. Nonetheless, the work schedule indicates the start of each project: the No.3 reactor from early 2015; and the No.1 and 2 reactors from late 2017[9]. Yet, neither the State nor TEPCO have any specific outline for the procedure.

Removal of spent nuclear fuel is an initial step of decommissioning. How much irradiation labour and time must be sacrificed in order to remove all spent nuclear fuel from the No.1-4 reactors? It is beyond my imagination.

---

[8] The operation started on 18 November 2013 and finished on 20 December 2014.

[9] According to the latest roadmap released in June 2013, the medium and long-term decommissioning operation schedule will be updated in spring 2015.

Decommissioning without vision

Melted fuel is impossible to remove

In order to decommission a nuclear plant, firstly all spent fuel needs to be eliminated completely, otherwise, there won't be a step two. The first step is now underway at the No.4 reactor. Including the No.4 reactor project, the difficult challenges are awaiting us in order to remove all spent fuel from all reactor pools. Even if one day in the future all spent fuel assemblies are removed, it will never be possible to remove melted fuel from the No.1 and No.3 reactors.

The state and TEPCO plan to remove them as soon as 2020. The agenda shows that the melted fuel will be extracted, but I don't think that is feasible. Because, I believe the melted fuel, as I explained the possibility of using metal to cool the reactor cores in chapter one, may be scattered and stuck here and there. There are probably more of them than just one lump sitting compactly in one place.

The total amount of uranium is 100 tonnes. Regardless of its weight and quantity, we don't even know where it has melted. Under such conditions, logically it is impossible to extract 100 out of 100. For example, if 50 of them are to be taken out, the other 50 will remain. Not only can the entire fuel not be removed, it also entails severe exposure of workers. Concerning these challenges, I doubt the task itself can be carried out.

Sarcophagi is the only option

In my opinion, the only option is to give up on the attempt of taking out the melted fuel, and enclose the entire reactor buildings with sarcophagus. That is probably the most realistic solution. Like in Chernobyl, the reactor buildings will be covered with concrete entirely. If the buildings are covered leaving nuclear fuel inside, it might lead to a radioactive leak by water absorption. Somehow, all nuclear fuel must be removed, but it may not be possible as it entails the radioactive exposure of workers.

As I repeat again and again, this is an accident that mankind has never encountered in the past. The worst case in the past was an accident in Chernobyl in the Soviet Union, where one fuel core melted down. In Fukushima, three nuclear cores melted down. In the No.4 reactor, the building is half destroyed, and the spent fuel pool that contains radioactive fallouts is hanging in the air. Such an unprecedented phenomenon is happening in Japan, and no one knows the best solution on how to decommission the nuclear plant.

Then, what is the priority?

My concern is the workers, including young people, who have no knowledge of irradiation, who are forced to be exposed everyday under the system of sub-contractors, and sub-sub-contractors, which goes down to even 10 sub-contractors. We should choose the best way to ensure the least irradiation to workers as possible.

Never-ending battle against radioactivity

I don't know how long it takes to construct a sarcophagus. Even if a sarcophagus is completed perfectly, it will crumble after 30 years, as is happening in Chernobyl today. In that case, second and third sarcophagi need to be constructed to cover it. Endless operations will continue several decades or several hundred years. Besides, all workers are exposed to radiation. It is beyond belief how this project can be accomplished, and how much tragedy is waiting ahead of us. In Chernobyl, 600,000 to 800,000 workers were involved in the decommissioning project. Twenty-nine years later, several thousand people are working there per day. Just one reactor requires such a number of workers. It is hard to estimate how many workers and years will be required in order to decommission the Fukushima NPP. Are there enough workforces in Japan for the decommissioning?

Perhaps I will not live to see the initial sarcophagi completed, let alone the second sarcophagi. Unfortunately, the battle against radioactivity continues with no end. It will continue many hundreds of years with workers' exposure to radiation.

It seems that the Fukushima accident is all over to most Japanese people. Mass media is playing a role for such movement. To me, it is insane.

What can we do with radioactive waste?

Tokai Nuclear Power Plant decommissioning plan delayed again

On 18 November 2013, the first day when the extraction of spent fuel assemblies began on the No.4 reactor at the Fukushima Daiichi NPP, the first day of decommissioning, the Japan Atomic Power Company (JAPC) announced the delay of the dismantling operation at the Tokai Nuclear Power Plant in the Tokai village (Ibaraki prefecture), which was originally planned for 2014. It was the second time that the Tokai Nuclear Power Plant dismantling operation had been postponed after that of 2010.

The Tokai Nuclear Power Plant started its operation in 1966 as the first commercial nuclear plant. The decision to close the nuclear power plant was due to its aging after 30 years of operation, and inefficient output of 166 MW with regard to the size of the institution. Since the operation was stopped in March 1998, the decommissioning process has been on going.

Normally, the decommissioning process is as shown in Figure 13 (page 86). In the case of the Tokai Nuclear Power Plant, they have reached the first step: stopping the reactors; the second step: extracting spent fuel; and decontaminating the equipment and pipes in order to reduce radioactivity (steps 3 and 4). Yet, they are stumbling at step 4 "dismantling and demolishing the interior of the buildings".

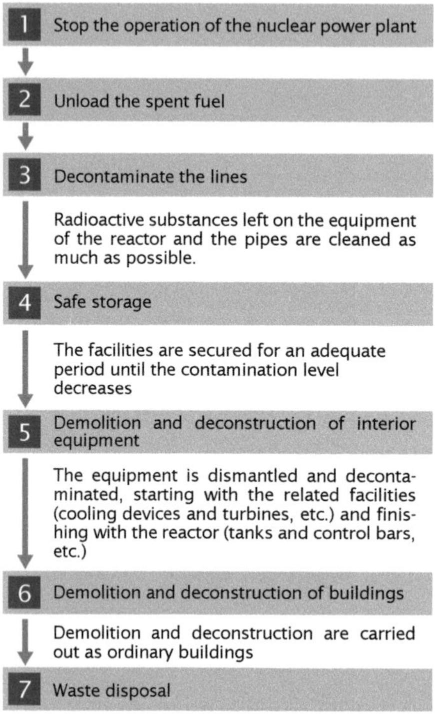

Source: Based on the diagram from the Asahi Shimbun Globe ("The decommissioning era", 21 July 2013)

Note: Several steps can be carried out simultaneously or in reversed order.

Figure 13: The steps to decommissioning

In the current situation in Japan, the decommissioning waste produced by dismantling a nuclear institution has to be buried and controlled as "low level radioactive waste". Low-level radioactive waste is classified into "extremely low waste", "relatively low waste" and "relatively high waste". Each rank has different operational standards. For a disposal site for "relatively low waste", there is a "low radioactive waste treatment centre" in Rokkasho Village, Aomori. But this centre only accepts nuclear waste generated from operating plants, not the waste from decommissioning plants. And the other two types: “extremely low waste” and “relatively high waste” don’t even have a place for storage. For the decommissioning waste, there are rules for disposal, but no place is ready for it.

The Tokai Nuclear Power Plant established its decommissioning plan with a notion that, since the rules exist, the disposal location would eventually be found. Then, the Ministry of Economy, Trade and Industry approved their plan. Even today, not a single location has been found. Currently, the Tokai Nuclear Power Plant is not the only one dealing with decommissioning operations. The Advanced Thermal Reactor Fugen in Atsuga, (Fukui prefecture) No.1 and No.2 reactors of the Hamaoka Nuclear Power Plant which is owned by the Chubu Electric Company in Omaezaki (Shizuoka-prefecture), and soon the No.4 reactor of the Fukushima Daiichi Nuclear Power Plant, if they succeed to extract all the spent fuel safely, are the ones that face a situation in the near future as “decommissioning, without capability”. As for the Fukushima Daiichi NPP, the No.5 and No.6 reactors are also on the decommissioning list, although they will not be dismantled, as they will be used as a training facility.

The concept of an “interim storage facility”

The waste from decommissioning is not the only waste that doesn’t have a disposal location. The most serious radioactive waste, spent fuel, and contaminated material assembled by decontamination operations don’t have final locations to be stored. Of course, the State wishes to decide on the destination of that waste as soon as possible, but nobody accepts such dangerous poison eternally. Thus, they started to look for options of an “interim storage facility”, where the nuclear waste would be stored temporarily. Nevertheless, no one wants to volunteer. Finally, for the contaminated waste assembled by decontamination operations within the Fukushima prefecture, which should be taken into consideration urgently, the State proposed nationalising the land around the Fukushima Daiichi NPP, and constructing the interim storage facilities.

The state will buy about 19 km^2 including the towns Futaba, Okuma, Tomioka, and Naeba. The state kept saying, *"people can go home once the decontamination is complete"*, then suddenly they announced this plan. It must have been a shock to the habitants, and I can imagine their vexation and resentment.

I deeply sympathise with habitants' emotions. Despite everything, with my 40 years career as a guardian of radioactivity, I must agree with the idea of centralising radioactivity in the most contaminated areas and blocking them from outside by concentrated operations. However, as I will explain further in a later chapter, disposal locations for radioactive waste should not be constructed in residential areas, but within TEPCO's land.

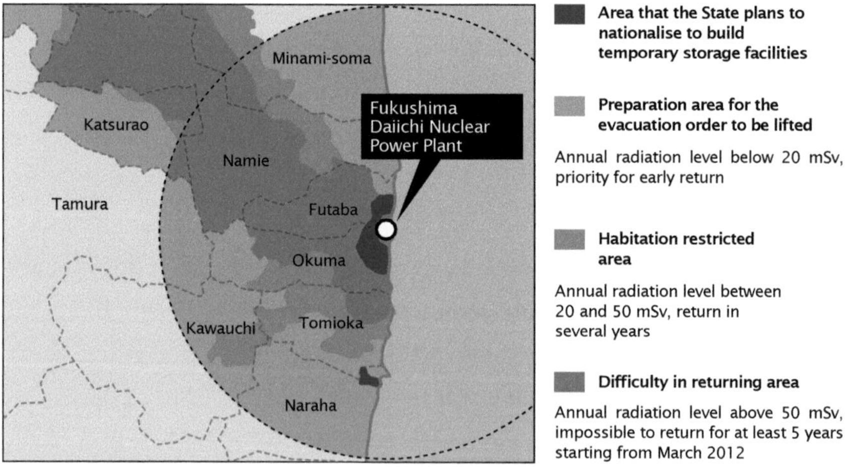

Source: Based on the map which appeared in the Asahi Shinbun ("The earthquake in the northeast of Japan, 1000 days later", 4 December 2013) and the Fukushima Minpo (15 December 2013)

Figure 14: The areas planned to be nationalised

Another aspect I would like to point out is about the promise the State made to the Fukushima prefecture. Upon the construction of interim storage facilities, the State said "*The state legally specifies that the contaminated soil will be ultimately treated outside of Fukushima within 30 years.*" Essentially, an interim storage facility is so named, because stored objects are planned to be sent to the final destination. Now, let us look at how interim storage facilities come to be a reality. Do you know the dream of pro-nuclear people called the "nuclear fuel cycle", the dream they cannot give up on today? I will come back to this dream in Chapter Five. To summarise, it is to extract plutonium (reprocessing) from consumed uranium in a reactor (spent nuclear fuel), then proliferate Plutonium in the special reactor called a high-speed breeder reactor, in order to support Japan's scarce resources with eternal electric resource. Hopefully, it might be possible to make nuclear arms with the Plutonium. That's their dream.

In the beginning, all spent fuel in nuclear power plants was planned to be taken to a reprocessing plant and treated progressively. The plan for an interim storage facility was not even on the table, or nobody had such an idea. However, as the reprocessing plant never started to work, more and more spent fuel was accumulated at nuclear power plants. With no other choice, the idea of constructing an "interim storage facility" was brought up for the first time in the nuclear world, in order to temporarily store spent fuel somewhere.

It was called an interim storage facility because all spent fuel stored there, a priori, was supposed to be transferred to a reprocessing plant.

An example of an interim storage facility would be the one in Mutsu, Aomori prefecture. This facility can store 5,000 tonnes of spent fuel. Although it is called interim, the maximum storage period is 50 years. The Rokkasho Village Reprocessing Plant, which is to be the final destination, is still under construction. Completion is scheduled for October 2014 but it is highly uncertain, as it has been postponed more than 20 times.[10] In addition, Monju, the high-speed breeder reactor, shows no signs of starting operation, which indicates that the nuclear fuel cycle has literally failed. Considering this fact, the spent fuel stored at the Mutsu interim storage facility will not be removed in the future, even if the Rokkasho Village Reprocessing Plant begins its operation, and the spent fuel in Mutsu won't increase its volume. In short, as I have been drawing attention to this point all this time, it means that the interim storage facility will become the definitive storage facility. In the case of Fukushima too, the State insists on the condition "interim" by saying the contaminated soil will be transferred to other prefectures in 30 years, which will be inscribed as law. However, as there is no definitive location, it is not logical that it be called interim. Above all, I have absolutely no faith in the Japanese Government, as they tend to break laws with no hesitation when the law they have made overwhelms the reality. Once it has been done, there won't be any changes later on, meaning that the contaminated objects and nuclear waste will stay there for good. Although it's called "treatment" and "disposal", radioactive substances cannot be eliminated by burning or by any other means. They won't disappear by being concentrated in the same spot. Basically, we can neither treat nor dispose of them. The only thing that can be done is to take care of the abandoned radioactive substances for a very long time.

The geological disposal of highly radioactive waste and spent fuel

It was 1942 when mankind first operated a nuclear reactor. It was when the United States created the atomic bomb. Immediately, they realised that a nuclear reactor produces radioactive substances called fission products, which is highly problematic if we cannot eliminate them. Therefore, they

---

10 Scheduled termination of construction is now postponed to March 2016.

have embarked on research ever since to eliminate the fission products. More than 70 years has passed with the hope that "one day, the problem will be resolved" and "humanity will discover the way to eliminate radioactive substances thanks to scientific progress". Unfortunately, mankind today still does not have the knowledge to eliminate radioactive substances being produced. In the first place, it is important not to produce any more radioactive substances that we have no capacity to clear up. Stopping the use of nuclear energy is more important than anything else, which is my forefront belief. Having said that, we must also deal with the fact that Japan has already produced radioactive waste equivalent to 1.2 million Hiroshima bombs.

Since they cannot be neutralised, the only option is to isolate them from the environment. The issue of how to isolate highly radioactive waste and spent fuel has been discussed worldwide. International treaties prohibited all proposals such as space disposal, ice sheet disposal in the South Pole, oceanic disposal. In such cases, the only option left is to store them somewhere on land. In order to store them correctly on land with the upmost technology, a container called a "dry cask" will be needed. However, setting up this facility costs so much that the promoters of nuclear energy in Japan chose to block radioactive waste inside of glass and bury it deep under the land surface, and they think it will be sufficient. It is called geological disposal. I don't think it is so simple.

How long is the estimated storage period of radioactive waste? They have to remain immobile during one hundred thousand to one million years and there is nowhere in this high earthquake-prone country to guarantee such condition. In fact, on 11 September 2011, the Science Council of Japan, which is sort of the parliament of researchers in Japan, announced that burying radioactive waste is not realistic.

Despite everything, the Japanese Government, who believe the geological disposal is the only possibility, cannot let go of this idea. In October

2013, the Ministry of Economy, Trade and Industry established a committee to re-evaluate the technological reliability of a geological disposal. Several months later, I found a shocking article in a magazine, which I quote as follows. First, I will quote the comment of Tadashi Narabayashi, the post-graduate professor in Hokkaido University, rejecting the comment of former Prime Minister Junichiro Koizumi, who was recently advocating zero nuclear energy. As Mr Koizumi claimed, *"Nuclear Power Plants should be stopped because we cannot opt for geological disposal"*, Mr Narabayashi explains the reason why he is against it:

> *[At the high-level radioactive waste disposal centre in Rokkasho] the solidified glasses containing dissolved radioactive substances are stored in stainless containers. They are sealed and lined up. It is planned they will be buried after being cooled for 40 years. In 40 years, the radioactivity and the calorific value will be reduced to around one thousandth, and they are safe to be buried. In reality, they have already been stored safely. (...)*
>
> *Solidified glass forms allow a storage period of one hundred thousand years. In other words, the fact that it can be stored for "one hundred thousand years"; does not mean the radioactive waste requires one hundred thousand years. Mr Koizumi doesn't know this fact nor do we already have the adequate technology. In 3000 years, the intensity of radioactivity will drop to one millionth, only 10 times more than uranium ore, which we can touch with our bare hands.*
>
> *(From the article "Koizumi's Zero-Nuclear Energy comment is fatally immature")*
>
> *Shukan Shincho (New Year's 2nd Special Edition, 9 January 2014).*

Mr Narabayashi is someone who, after the Fukushima Daiichi NPP accident, advocated, *"The accident would be under control, if I were in charge"*.

I was appalled by his continuance of lies. First of all, what is referred to as *"solidified glasses containing dissolved radioactive substances"* is the fission product, from which uranium and plutonium had been removed from spent fuel after being treated in France. Among them, there are substances whose half-life is relatively short, thus it is true that the radioactivity will be reduced to one thousandth in 40 years. However, even one-thousandth of radioactivity can easily kill people if they go near it. It is impossible to bury it safely.

Furthermore, is there a technology to store the solidified glasses? Unfortunately, no science or technology can guarantee safety for 100 thousand years. That is the reason why the Science Council of Japan rejected geological disposal.

Now, let's look at Mr Tetsuo Sawada's (associate professor at the Tokyo Institute of Technology) comment:

> *"Recently, I had a chance to talk about the radioactive waste issue to junior high-school students. I introduced the fact that if we distributed waste from nuclear plants to each individual in Japan, it would only be the size of a handful for each person, but we have no place to dispose of it. The students proposed to dispose of it in the ocean, dispose of it in space, or dispose of it under the seabed. Otherwise, each household could take care of it. Although the last proposal is unrealistic, it leads us to an idea to manage radioactive waste collectively by accepting it in the capital. This sense of public well-being shown by these junior high-school students can be an answer to the problems."*
>
> *(From the article "Koizumi's Zero-Nuclear Energy comment is fatally immature")*
>
> *Shukan Shincho (New Year's 2nd Special Edition, 9 January 2014).*

This was probably from a visit lecture at a junior high school. I agree with the idea of accepting nuclear waste in the capital. However, when he explained that nuclear waste is only a handful in size, he should have also explained that the handful of poison could kill them instantly as soon as it's placed on their hands. Otherwise, the accurate knowledge about radioactivity cannot be passed onto the children. With such an insufficient explanation, I cannot deny that it is an act of brainwashing.

Mr Nakabayashi and Mr Sawada have been promoting nuclear energy as *"nuclear energy is safe"* and *"no catastrophic accident will occur"*. They have diffused unpardonable false information and knowledge. Although the Fukushima Daiichi NPP proved their fallacy, they keep claiming that nuclear waste can be treated easily, without recognising the hardship of the Fukushima people. I feel shame towards them as a scientist.

Then, what can we do about the storage of highly radioactive waste and spent fuel? In my opinion, it is relatively rational to keep them on the surface, such as in temporary storage facilities with dry casks. The economic effect should be ignored at this point. No matter how much it costs, we should take responsibility in order to prevent harm to future generations. Having said that, there is no absolute, concrete and safe method. We have now come to an inescapable reality. Among these choices, the relatively reliable one is to continue storing them on the ground where we can at least keep our eye on them. That is the only choice to be made.

The criteria for choosing a final disposal site

In principle, disposal waste should be returned to its owner

The Japanese Government is desperately looking for locations to bury and dispose of nuclear waste. In order to construct the facilities to store the radioactive waste generated by the decontamination operations in the Fukushima prefecture, there is a movement to nationalise the land around the Fukushima Daiichi Nuclear Power Plant. Also, the State had a hard time finding candidates by means of a public appeal system to construct the final disposal sites, so the State changed to a proactive policy to designate certain areas as potential disposal sites. At first, the State undertook the selection of locations for a final disposal site for the "designated waste". The designated waste came from the contamination by the Fukushima Daiichi NPP accident, and is currently stored temporarily in eleven prefectures. This is the radioactive waste exceeding 8000 Bq/kilogram. The total amount, according to Asahi Shinbun on 20 January 2014, is about 132,000 tonnes as at the end of August 2013.

The Ministry of the Environment has a plan to construct the final disposal sites in five prefectures: Miyagi, Ibaragi, Tochigi, Gunma, and Chiba. In January 2014, the nationalised lands in three cities and towns of Miyagi prefecture were put forward as the first candidates. Among the nationalised and prefecture-owned lands, potential locations where the final disposal facilities could be constructed were selected and went through inspection. The distance from residential areas and water sources, and the level of environmental protection were measured and evaluated in order to select the ideal candidates. However, the strong resistance of local leaders was inevitable. Despite such resistance, Mr Yoshihiro Murai, the Governor of the Miyagi prefecture made a decision to accept highly contaminated radioactive waste at one of the three candidates in Miyagi, while sympathising with local leaders' resistance as *"it is a natural reaction as representatives of cities and towns"* (NHK news web, 20 January 2014),

In my opinion, the radioactive waste spread by the Fukushima accident should not be accepted by any communities or the administrations. Let me explain to you my principal rule on the treatment of radioactive waste.

Radioactive waste must be treated: 1) on site, where it is produced, 2) made as compact as possible, 3) confined, without being mixed with other waste.

The radioactive waste has already spread to wide areas, and it is impossible to confine it in all contaminated areas. In such a case, it should be gathered in a proper location and confined. Where would that location be? The accident occurred at the Fukushima Daiichi Nuclear Power Plant and all the radioactive substances scattered everywhere originally came from there. Therefore, they should be returned there. The source of contamination is originally the radioactive substances in the reactor cores at the Fukushima Daiichi NPP owned by TEPCO, which makes TEPCO the owner of the radioactive waste. It is logical to return the property to its owner. Normally, it should be returned to inside the Fukushima Daiichi NPP, but it is not possible to add the volume of radioactivity to where many workers are currently fighting against radioactivity with their lives at stake. Accordingly, I suggest using the Fukushima Dai-ni (No.2) Nuclear Power Plant, which extends to the south of Fukushima Daiichi, as a radioactive disposal site. In order to prevent the contaminated materials from spreading, I strongly hope they will choose this method instead of accepting the waste in many municipalities in Japan.

To my surprise, TEPCO announced a plan to prepare to restart the Fukushima Dai-ni Nuclear Power Plant, while they have not resolved the accident at Fukushima Daiichi at all. I don't think it is justified to run away from the problems without being condemned, and even restart another nuclear plant, while compelling distress to millions of people around the site. Rather, the Fukushima Dai-ni Nuclear Power Plant should be sacrificed first as a disposal site. If it's not large enough, TEPCO should offer another broad site – the Kashiwazaki-Kariwa Nuclear Power Plant.

Public appeal system at a public meeting

Hypothetically, let's say that the final disposal site issue can be dealt with by my former proposal, but radioactive waste at a volume of 1,200,000 Hiroshima bombs still exists in Japan. It is inevitable that other disposal sites must be found in order to deal with that waste. As I explained, the State has changed the policy to designate the potential locations, but I disagree with this approach. It should remain within a public appeal system. The existing system until now seemed democratic, but in reality, the candidates who put themselves forward were the localities that were financially impoverished. This is not the way decisions are supposed to be made. Then how can we decide? First, it is not fair that one particular group monopolises the information on nuclear energy and influences the public with one-sided arguments. If one locality puts itself forward as a candidate to accept the waste, the candidate deserves to hear opinions from the anti-nuclear side as well as the pro-nuclear side.

The simple public appeal system is not good enough. It is also important to not just depend on localities to voluntarily put their names forward, but also include specialists to provide academic overviews when choosing the locations. I still believe that stocking waste on the surface is better than geological disposal. However, if the State pursues a geological disposal method, the first thing that has to be done is to look into which location in Japan has the best conditions to bear high-level nuclear waste and spent fuel underground. Studies from multiple perspectives should be undertaken and proven to the public.

Moreover, the idea of designating Tokyo or Osaka as candidates, where the habitants consume massive amounts of energy, must also be examined from a social point of view. Then everything must be discussed with transparency. Traditionally, transparency didn't exist, and the decision was supported by nuclear promoters' propaganda such as: *"Nuclear energy is safe", "To accept [the nuclear waste] is honourable",* and *"it is good for environmental protection".*

It is obvious that the treatment of high-level nuclear disposal and spent fuel entails risks. No matter what methods are used, whether it is surface disposal or geological disposal, we cannot avoid leaving the debt to future generations. To date, the State and the pro-nuclear organizations imposed everything on depopulated regions, with a one-sided explanation, in exchange for money. At the very least, such a biased way should be stopped.

How to protect ourselves from radioactivity: Q&A

Food

Can we measure Becquerel at home on vegetables and other food? Is there any visual way of detecting it?

I am sorry to say this, but my answer is "No". I am sure many of you would like to measure the radioactive doses, and may have bought devices to measure radioactivity. Fundamentally, it is not possible to measure the level of radioactivity with such devices. Measuring radioactive amounts is extremely difficult. It is better not to think that radioactivity is measureable by ordinary people. It has to be measured by experts at a designated institution.

I went to Minami Soma city in the Fukushima prefecture two years after the accident at the Fukushima Daiichi Nuclear Plant. In the city, people continue with their daily lives, just like before.

In fact, the radioactivity in that area is so high that it is supposed to be designated a "radiation controlled area". However, even under these conditions, people, including children, continue living ordinary lives.

Human beings are programmed by genetic information and DNA, which construct information, are connected with molecular bonds with the energy of few electron volts. Compared to energy bonding DNA, radiation such as the energy generated by gamma rays from cesium 137, is 661,000 electron volts: hundreds of thousands times more.

You have all experienced your temperature rise by one degree when you have caught a fever. No one dies from that. But if you receive energy from radiation, your genetic information chain will be torn asunder by just a temperature increase of two thousandths of a degree. Humans would die 100%. Radiation cannot be felt with our five senses. What we call "radiation" is actually "radioactive substances". They are substances with forms and colours, even smell.

However, because radiation is so predominantly poisoning to human bodies, we wouldn't be able to survive if there was enough radiation for us to feel. The total amount of cesium 137 defused in the air at the accident of the Fukushima NPP was only 5 kilograms. Only 1 kilogram of cesium 137 has fallen on the land of the Tohoku and Kanto areas. Obviously, it is invisible. What we are trying to measure are such elements that it is best not to deem as measurable.

School lunches: Are they measured correctly?

The Government and councils are responsible for measuring school lunches correctly. If you have a doubt, you should ask them and confirm directly. In order to measure radioactivity in food, each product must be screened independently and examined one by one (rice, meat, vegetables, cucumbers, tomatoes, carrots, spinach etc.). The data should be accumulated for each product to reflect the accurate procedure.

Also, the workers involved with school lunches should be aware of the importance of professional responsibility to reduce the irradiation of children. Even if they are aware of it, it may be difficult to protect children entirely, while the State advertises that 100 Bq/year is safe and municipalities are in charge of the program. At the same time, those workers should be aware of the basic knowledge of how to protect children, who are extremely sensitive to radiation because school lunches are provided to growing children whose cell division is highly active.

Children should eat less contaminated food. What is the standard?

When we set a standard, people tend to believe that anything under that standard is safe. That's why I am against setting a standard.

The state affirmed that less than 100 Bq/kilogram of cesium is safe. But a risk exists at even less than 100 Becquerel, or 99, 90, 50, or 10 Bq/kilogram. Before the accident at Fukushima, most of the food in Japan contained less than 0.1 Bq/kilogram of cesium.

If we accept 10 Bq/kilogram, although this is clearly better than 100 Bq/kilogram, it means 100 times more contamination is accepted compared to before the accident. I definitely do not want food that contains 10 Becquerel to be given to infants and babies. It is hard to respond to the question on how much radioactivity we can tolerate. I would say, 0.1 or 0.5 Becquerel, the amount as close to the amount prior to the accident. It is normal that people feel unsafe as the State tolerates 1,000 times more contamination compared to before the accident. Some criticise me saying I am spreading a harmful rumour, but unfortunately, it is not a rumour but a real danger.

Domestic equipment and environment

Are the radioactive substances removed by water filters?

Figure 15 (page 103) graphically shows the measurements of the atmospheric concentration of radioactive iodine 131 measured by myself at the Research Institute of nuclear reactors Kyoto University, located in Kumatori (Osaka Prefecture) after the Chernobyl accident in April 1986. The black dotted line at the bottom shows the data obtained with a bare paper filter and the top line, consisting of squares, are the numbers obtained with a carbon filter (activated charcoal). The reason why the levels drop day by day, is because the half-life of radioactive iodine 131 is as short as eight days. A comparison of the approximate values of these two lines allows us to understand that the carbon filter absorbs five times more iodine than paper. The activated charcoal thus possesses an absorbent capacity applicable to various substances, including iodine, for which it is used in effluent disposal systems and for cooling nuclear power plants.

Hoping that a water filter with activated carbon can partially eliminate radioactive iodine, I conducted an experiment, in particular, at the house where the contamination of breast milk was detected immediately after the Chernobyl accident. I installed activated carbon filters to a water tap, in order to obtain clean water by eliminating radioactive iodine, hoping to cleanse the body of the mother, whilst at the same time to get clean water to prepare powdered milk bottles. After about a month of using filters, I collected the filter to measure radioactive substances, but unfortunately, despite my hope, the active carbon had shown no effectiveness.

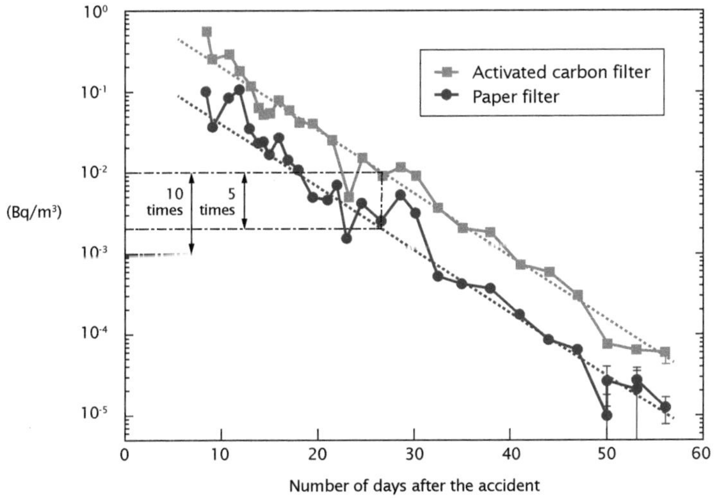

Figure 15: Atmospheric concentration of radioactive iodine 131 measured at the Research Institute on nuclear reactors after the Chernobyl accident

We live in an area where the level of radioactivity is quite high. What can we do to protect our children?

There are two types of irradiation: internal and external. We have to protect ourselves from both.

First, let's look at external irradiation. External irradiation means being exposed to radiation by staying within the place where the radiation level is high, or touching contaminated objects. Therefore, the places where children spend a lot of time must be decontaminated, for example, schools. The top layer of soil in their playground should be removed and transferred to somewhere else. The soil at the playschool should also be removed, as well as in parks and the backyards of individual houses if children play with mud. Of course the "black particles", which I mentioned in the previous chapter, should be carefully removed. This project cannot be carried out on an individual level. Moreover, in such highly contaminated areas, it is important to think with collective perspectives, instead of only worrying about your own children. In this respect, the Government should be the ones to take charge, and the people should ask for this.

In addition, internal irradiation must be taken into consideration. Internal irradiation occurs by two means: by respiration and by food and drink. The only way to prevent internal irradiation by respiration is to wear a mask. The period during which radioactive substances were floating in the air was for two weeks after the accident. It was understandable at that time, but wearing masks every day is somewhat suffocating. I cannot imagine children running around with masks on. In the recent situation, it may not be necessary. However, the radioactive substances once fallen to the ground may be blown up in the air on windy days and in sandy places. Masks are not as effective as is believed when dealing with blocking radioactive substances. The masks that fit perfectly on the face are expensive and make it difficult to breathe. Ordinary masks have space for aeration, which means no efficiency.

The other means of internal irradiation is through food and drink. Needless to say, we have to choose the least contaminated products as possible. At home and for school lunches, the products must be correctly controlled. For the moment, this is the most important task to be taken care of.

How to spend everyday life

We don't live in a contaminated area. Should we be careful about getting wet by rain?

I would like to say it's all right now, but the expression "all right" should not be used here. When it comes to radioactivity, there is no such thing as "secured" and "all right". Even a very small amount of irradiation may cause negative effects. So if I were asked if getting wet by rain is all right or not, I would recommend not getting wet. Four years after the accident at Fukushima, the atmosphere in Japan, with the exception of Tohoku, the north of Kanto and some parts of Tokyo are almost not contaminated at all, at least not by the emission from Fukushima. As for rain, I would not worry too much about this, but it's up to you to decide.

I would like to take my children to the beach. Should it be avoided?

Before the accident at Fukushima, people went to beaches to swim without giving it a second thought. However, the seawater was already polluted at that time. In the 1950-60s, the United States and the Soviet Union carried out atmospheric nuclear tests. Nuclear and hydrogen bombs exploded many times in the air.

Regarding the estimated fallout of radioactivity in the northern hemisphere, where Japan is located, the total amount was 5,000 Bq/m^2. In 1985, the Chernobyl Nuclear Power Plant had an accident. The fallout on Japan caused by this accident, although it depends on the regions, was approximately 100 Bq/m^2. With the 100 Bq/m^2 fallout, food products got contaminated here and there, which, for example, caused the suspension of green tea leaves from being shipped.

Today, the fallout of the Fukushima accident is added on top of that. How much would that be? Around the Fukushima Daiichi Nuclear Power Plant, it's about 600,000 Bq/m^2. From the Fukushima Daiichi NPP, probably about fifteen thousand square kilometres of radiation controlled areas exceeding 40,000 Bq/m^2 is spreading over the Tohoku and Kanto areas.

This means the ocean near the Tohoku and Kanto regions is most likely contaminated and as seawater is all connected, the Pacific Ocean and the inshore of Japan may also be affected, maybe not be as bad as to be designated a "radiation controlled area", but its contamination level is probably equivalent to the impact of contamination by atmospheric nuclear testing.

Other mothers around me do not talk about radioactivity anymore. Am I obsessed as I worry about it even today?

After four years have passed, everyone starts to forget as if the Fukushima accident never happened. It is quite normal as the State and TEPCO are controlling information intentionally, to make people forget about it. Unfortunately, the accident at the Fukushima Daiichi Nuclear Power Plant is not at all under control. At the site, many workers are exposed to radiation while working hard every day in order to prevent the radioactive substances from being released. The radioactive fallout has already been emitted to the lands and the ocean, then crops are harvested from the contaminated lands. We should not forget it. In particular, it is an adult's responsibility to protect children from irradiation, so we should keep an eye on children's safety.

Nobody likes to live in fear, and everyone wants to believe that the situation is safe. It is natural that people want to forget, and people will forget it even more as time passes. However, please don't accept the reality meekly. If we accept the situation as it is, it will lead to further damage in the future. Besides, children have no responsibility for this accident. Hence, it is again the responsibility of adults, and I hope adults will continue making the effort to provide less contaminated food to children.

The impact on our bodies

My child sometimes has nosebleeds for no reason. Is it because of radioactivity?

In terms of the impact of irradiation on human bodies, one clear effect is that a great amount of irradiation causes death. After the atomic bombs were dropped in Japan, numerous research was conducted to examine the impact of non-fatal irradiation. One primary result was that the number of leukaemia patients increased. Also, the number of cancer patients increased. Recently, research shows the number of cardio-related diseases has also been increasing. Progressively, the research data started to prove that other diseases and symptoms have a higher tendency to occur more among the irradiated population. To summarise, after 68 years (as at February 2014) since the atomic bombs were dropped, we have started to understand that the number of diseases are increasing. This is our history.

As explained previously, radioactivity has an overwhelming level of energy compared to that of humans to live. Thus, anything could happen, not only leukaemia, but also cancer, and cardio-diseases are caused by radioactivity. Nosebleeds could be a symptom. Unfortunately, researchers cannot scientifically prove the facts with feasible data that those results are related to radioactivity. It will remain difficult to obtain such data.

The only thing I can say is that if your child may have been exposed to radiation, you should be aware that anything could happen, and to keep an eye on your child. Paying attention not only in terms of choosing food and drink, but also by getting rid of mud after playing, limiting swimming time at the beach, and measuring the radioactivity level when you go into the mountains. In order to reduce children's exposure to radiation, adults should take all measures possible.

Do thyroid cancers of children in Fukushima have any relation to the accident at the Nuclear Power Plant?

In Fukushima, 33 children have been diagnosed with thyroid cancer, and 75 with suspected cases, to date. (Released 7 February 2014)[11]

---

[11] The last report dated 24 August 2014 reported a total of 103 confirmed or suspected cases of thyroid cancer among children in the Fukushima Prefecture, 14 more than three months earlier.

The number of children tested was 254,000, which makes the crisis rate (with the 33 children who have been diagnosed) 13 out of 100,000 children; 130 out of 1 million children. The international standard crisis rate of thyroid cancer among children is 1 or 2 out of 1 million; therefore, the data above is showing phenomenal numbers.

Including this report, the prefectural health control management research investigation committee has been claiming, since the initial discovery of thyroid cancer, that *"there is no relation between children's thyroid cancer and the irradiation from the accident at the Fukushima Daiichi NPP"*. Such a pity... Science is a study supported with "doubt". By investigating the facts taking place before our eyes, scientists must take a critical view and examine each phenomenon carefully. The investigation committee wanted to believe nothing had happened in Fukushima, so they believed it. Based on such a belief, an investigation plan was established and led to this conclusion. They raised several reasons why the data is not influenced by irradiation from the Fukushima accident. First, the claim *"thyroid cancers were only found 4 years after the case of Chernobyl"* is a mistake. In Chernobyl, clear evidence of cancer began to increase after four years. Thyroid cancer already existed one year after the accident. Then, after observing an extraordinary crisis rate today compared to before, they said it's thanks to the improvement of test techniques. It also indicates that they have never tested it under the same conditions. If they have no knowledge of the past tests, they should say, *"We don't know"*. Obviously, something unusual is being observed among the children in Fukushima where the nuclear accident took place. If that's so, they should suspect the cause of the phenomena and investigate it thoroughly as they may be related to irradiation. That is the scientific approach. Under such limited conditions, I regret to say that I cannot affirm if this thyroid cancer is as a result of irradiation or not.

Simple questions

You have been working at an institution where the radiation level is high. How come you don't have any health problems?

I am not sure whether I have no health problem at all. My primary work is to measure the radioactive concentration level of various samples. It does not mean I am staying in a radiation controlled area from morning to night. When I am in a radiation controlled area, I do my best not to be exposed to radiation. Therefore, I am not irradiated as much as people think.

How am I being careful when I am in the radiation controlled area?

It depends on the work, but when necessary, I wear a mask and a protective suit, otherwise I avoid places where the radioactivity is high. A radiation controlled area is an area where the radioactive level is already high. Inside, there are zones with high radioactive levels and zones with low radioactive levels, and I know exactly where they are.

Having said that, it is true that radiation is invisible and insensible from our five senses. How can we detect it? We measure the quantity of radioactivity in each area with a dosimeter. I carry a dosimeter with me, but the dosimeter only tells me the level of radioactivity when I am physically in the spot, which is too late. It is better to know it before stepping into that zone.

The other day, the Government proposed a policy to people who returned to their homes, to keep track of one's own irradiation level by carrying a dosimeter. Ideally, the contaminated areas in Fukushima such as the cities, towns and villages should be measured precisely and the people should be informed of the level of radioactivity beforehand, instead of by just distributing dosimeters.

However, unlike my work environment, the areas are so vast, it is impossible to physically learn which areas are contaminated. In order to inform the people more effectively, how about using electric poles, for example? This way, even those who step in the place for the first time can see it clearly. On the electric poles, names and numbers of streets are already indicated. How about posting the rate of micro-sievert/hour on them? That will enable habitants to measure the dangers in each area, so they can choose their actions accordingly.

Do you apply any special technique in order to remove radioactive substances from your body?

No. As I am working in a radiation controlled area, I am exposed to a risk of inhalation of radioactive substances. Once they are in your body, fundamentally, it is impossible to remove them artificially. On the other hand, I should say, humans eat various things every day and do excrete them one after another. This condition is the same for radioactive substances.

However, it is basically not possible to speed up this process.

Exceptionally, if great quantities of radioactive substances are taken into a body, there is a medical treatment to force them out, but it will cause severe side effects. In ordinary cases, it is better to understand that there is no way to remove radioactive substances artificially.

Why don't the State and TEPCO disclose accurate information?

It is true that the Government and TEPCO are not sources for accurate information. In fact, historically, information has always been treated like that. The ones who have the authority disseminate only the information that is convenient to them.

Sometime ago, I read in a magazine a dialogue between Ms. Yuzuki Muroi, a writer and Mr Hirohiko Izumida, the Governor of Niigata.

Muroi: (regarding restarting the Kashiwazaki-Kariwa Nuclear Power Plant) You said, "Before the restart, the Fukushima Daiichi NPP accident must be investigated." *This statement is clear and direct.*

Izumida: But when my comments came out on TV and in the newspapers, that part is edited. (...) They write something like "Izumida is angry", *and the contents are ignored. My image will be controlled as I am being emotional."*

Shukan Asahi (issue 20 December 2013)
"The inconvenient truth" of nuclear energy concealed by the
Secrets Protection Law

In the same article, Ms. Muroi explained that she couldn't make negative comments about nuclear plants on television broadcasting and in newspapers. After the accident, she was restricted for using words such as *"critical"* or *"meltdown"*. She continued expressing comments such as *"I am scared"* and *"[the information of the State] is hard to believe"* as a comment on a personal level. Then, a businessman she passed on the street said to her *"You should die."* I will come back to this point later, but this situation is similar to the previous war when the information could only be heard from the imperial general side.

At the same time, the Japanese Government was sending accurate information to the United States. After the accident, the information about SPEEDI was not available to Japanese citizens as the information of SPEEDI wasn't accurate enough. Actually, SPEEDI was functioning correctly, and it announced the information such as which directions were dangerous on what day and at what time. And the information was sent to the United States. That's why the US soldiers left a distance of 80 km, immediately before the radioactive substances arrived in their direction, by carefully following the data. It's not just about SPEEDI. Once the information is sent to the United States, it is more or less spread to the world, so the rest of the world feels pity for the Japanese people.

As long as we stay passive, we won't have access to accurate information. If you have noticed that the Government and TEPCO are not providing appropriate information, you should actively search for the valid information. If you are able to think that way, Internet (whether it is reliable or not, that's another story) can be one medium for accumulating information, then you should obtain the ability to judge which information is correct.

Should we continue using nuclear energy despite everything?

We have sufficient electricity with zero nuclear energy

It is good to economise energy, however...

Some people believe that we will run out of electricity if nuclear power plants disappear. As a matter of fact, it has been proven that because of the Government's and TEPCO's threats, many people are trying hard to reduce the use of electricity and sufficient energy is being generated without using nuclear power plants.. Those people might believe that they can overcome hot summers and cold winters without nuclear energy thanks to their efforts. On the other hand, some might rather want nuclear energy back in order to avoid such an inconvenient lifestyle. It is a good thing to save electricity, and I am not discouraging it at all. However, I just wanted to make it clear that even without saving electricity, or without relying on nuclear energy, there has always been sufficient electricity.

During the last 80 years, there has been a shortage of energy for only several hours in peak summer

In order to prove this, I have created a graph, as shown below (Figure 16, page 114).

The graph shows the evolution of the installation capacity of electricity production, and the maximum demand for electricity from 1930 to 2011.

The bars indicate the installation capacity of electricity production from the bottom: hydroelectric power generation, thermal power generation, nuclear power generation and private power generating facilities. They represent the potential of the total production of electricity. The white dots indicate the maximum electricity demand, which means, the quantity of energy consumption measured at its highest day of the year. From the beginning to the mid-1990s, a few dots dropped in the area of nuclear energy, but most of the time hydroelectric and thermal power generations have been sufficient. Even in the years when there was a shortage of electricity, it only lasted several hours during peak-summer days. In those cases, private power-generating facilities served as a back up to cover the needs. Even now, certain members of the National Diet announce that at the scene they are walking a tightrope in order to avoid blackouts, but I think such a comment does not make any sense.

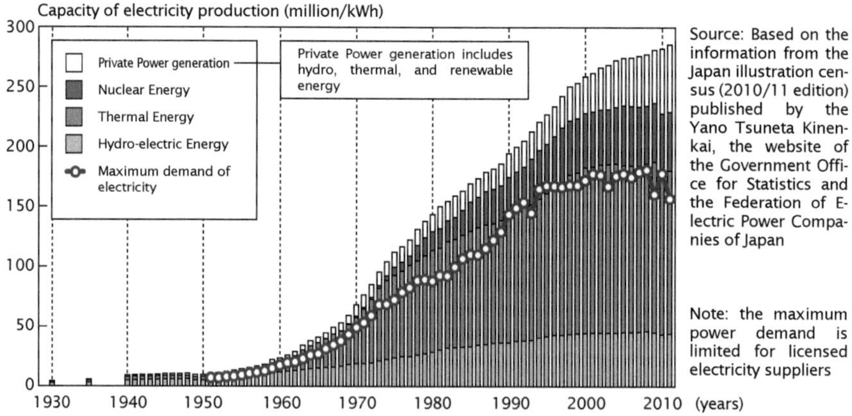

Figure 16: Evolution of the installation capacity of electricity production and the maximum demand for electricity

The cost of nuclear energy pushes up electricity prices

Nuclear energy involves such a huge cost

The electric power companies blame thermal power generation plants for the high cost of fuel, as a reason for electricity price increases. The Prime Minister Abe commented that about four trillion yen of the national budget is paid out to foreign countries due to fuel expenses. A statement like this may mislead people into thinking that thermal power generation is more expensive than nuclear power generation, but that is not true.

Several years ago, Kenichi Oshima at the Ritsumeikan University calculated the unit price of each energy source, based on the actual management data of electric power companies as well as their financial security reports. It turned out that nuclear energy was the most costly source.

(unit: ¥/kWh)

	Direct cost for generating electricity	Strategic cost		Total
		R&D cost	Geographical installation cost	
Nuclear Energy	8.53	1.46	0.26	10.25
Thermal Energy	9.87	0.01	0.03	9.91
Hydro-electronic Energy	7.09	0.08	0.02	7.19
General Hydro-electronic	3.86	0.04	0.01	3.91
Pumping-Up	52.04	0.86	0.16	53.07

Source: Extracted from the "Cost of Nuclear Energy", Kenichi Ojima (Iwanami Shinsho)

Table 7: The cost of a nuclear plant calculated by Professor Oshima (average cost between FY 1970 and FY 2010)

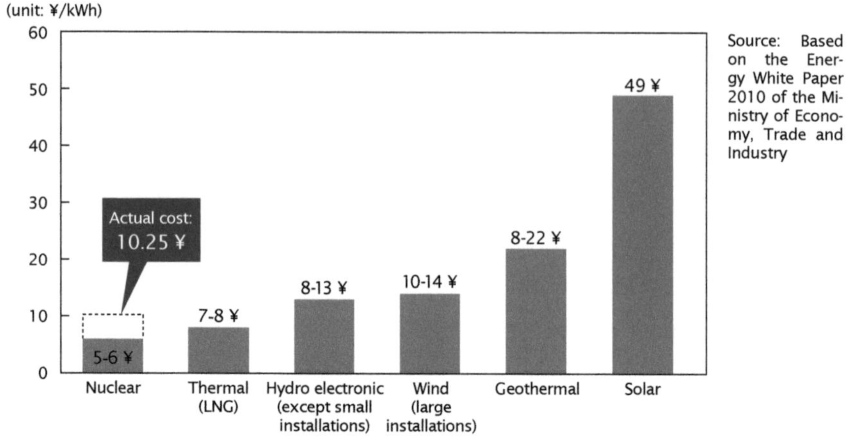

Figure 17: How the Government promotes nuclear energy

Let me show you why nuclear energy costs so much. First of all, let us compare the differences of plant construction fees for each energy source.

- Tokyo electric power company, Kazunogawa hydroelectric pumping-up power plant (1600MW): around 380 billion yen.
- Hokkaido electric power company, Tomari nuclear plant No.3 reactor (912MW): around 300 billion yen.
- Tokyo electric power company, Futsu thermal plant (natural gas), trench No.3 (1520MW): around 170 billion yen.
- Hokuriku electric power company, Atsuga thermal plant (carbon), reactor No.2 (700MW): around 127.5 billion yen.

Each electric power company disclosed the data above. The most expensive one is hydroelectric energy, followed by nuclear energy. But if they are calculated on an output voltage scale, nuclear energy is more expensive. Next is the expense for fuel. The promoters of nuclear energy often point out that: *"The energy generated by 1 gram of uranium is equivalent to that of 3 tonnes of carbon and 2,000 litres of petrol."* When you hear a report like this, it is easy to believe that nuclear energy, which is generated by uranium, is efficient. In fact, this theory is misleading. Petrol and carbon extracted from the ground are totally combustible as fuel. Among extracted uranium, on the other hand, 99.3% is non-fission (238). In order to collect 0.7% of fission uranium (235), the process required is called "uranium enrichment", which consumes an enormous amount of energy and entails huge costs.

Furthermore, nuclear energy involves significant amounts of money such as subsidies for the habitants at the sites, donations and reception fees for mass media (newspapers and television), politicians and bureaucrats, and universities. This "nuclear money" is outrageous. Besides, because nuclear plants cannot be constructed in big cities, they are situated in depopulated areas, which makes the transmission of electricity more expensive.

In addition, the cost for the re-processing of spent fuel and the decommissioning of nuclear plants, which doesn't exist for hydroelectric and thermal energy, should not be forgotten. Once an accident happens, as is the case with TEPCO today, an immensely large sum of money (environmental cost) needs to be paid for compensation and the decontamination processes.

As a matter of fact, there is an additional reason for high costs of nuclear energy, which is "pumping-up power plants".

Nuclear energy, unlike other energy sources such as hydroelectric and thermal energy, is not capable of easily adjusting the output. The level of output remains consistent regardless of the demand for electricity. But the produced electricity that exceeds the demand cannot be stocked. While a nuclear power plant keeps running during nights when less electricity is needed, redundant electricity is inevitably produced. In order to get rid of too much electricity being produced, a gigantic generating system is invented, with which water is transferred between the dams above and below. Generating electricity by discarding electricity sounds strange, but this is how the mechanism works. When too much electricity is produced, the water is transferred from the dam below to the dam above by using the surplus electricity. And when there is a shortage of electricity, the water jets down from the upper dam to the lower dam in order to generate electricity. It may seem like an intelligent system, in fact, it is an awkward mechanism as more than 30% of energy is lost when the water is being transferred. These electricity disposal facilities are installed at every nuclear power plant in Japan. The cost for constructing each system is estimated at approximately 150 to 200 billion yen.

Lax management of electric power companies is caused by the "fully distributed cost method"

As mentioned previously, electricity charges in Japan are calculated on the basis of the costs including generating electricity and policymaking, environmental costs, sales oriented costs such as PR, advertising, and donations, as well as personnel and management costs. All of the costs are added to the values of the owned assets, and calculated with margins on top. Examples of assets include not only institutions such as nuclear power plants, hospitals, recreational facilities, but also spent fuels and re-processed plutonium. This system is known as the "fully distributed cost method". All electric power companies standardise electricity charges as stipulated in the Electric Utility Industry Law. What this means is that the electric power companies are guaranteed their costs to be refunded from the electricity charges. The law protects their profit from the start. It's not only their profit that is protected, the more money they spend (i.e. constructing a new nuclear power plant and more PR campaigns), the more profit the system enables them to make.

Ordinary companies try to minimise the costs for facilities by hiring less expensive contractors in order to maximise on profit. Electric power companies are the other way around. The more money they spend, the more profit they receive, that's how it works. Thus, electric power companies look for constructors that cost 20-30% higher than others. Spending enormous amounts for publicity expenses, they spread untruth in order to produce the costly electricity, they then try to sell the electricity at even higher prices. The people who crowd around wanting a piece of the profit support the mechanism. As a result, we pay the highest electricity charges in the world. If the fully distributed cost method continues, and the current lax management is allowed to continue, through which anything they do will generate profits, our electricity charges will increase even higher in the future.

Originally, if we had never started using nuclear energy, we could have stayed with much cheaper electricity charges. Unfortunately, with the poor management of electric power companies, we have been paying expensive electricity charges. Until today, I have been protesting that nuclear energy be stopped immediately and all the existing nuclear power plants be decommissioned. If this idea is accepted, it will cost a huge amount of money for the decommissioning. If nuclear energy is abandoned, the existing nuclear plants will be left as bad debt. Then consumers must cover the payment for the debt, which means even higher electricity charges. However, if we keep up with paying the high electricity charges, the debt will only accumulate like a snowball. No matter how bitter the reality is, a decision must be made as there is no choice but to pay off the debt accumulated to date. If we abandon nuclear energy, which is unconditionally expensive, and change to hydroelectric or thermal energy, the electricity charges will decrease.

Nuclear energy is not clean energy

Nuclear power plants produce CO2 by generating electricity

A lot of people believe that nuclear energy doesn't produce carbon dioxide at all. As a matter of fact, a huge amount of carbon dioxide (CO_2) is emitted into the air when nuclear power plants are in operation.

Fission reaction itself doesn't produce any CO_2. In order to produce uranium fuel, uranium needs to be extracted from mines, refined, and concentrated. Then fission uranium is assembled and converted into a combustible condition in a reactor. In addition, the facilities must be constructed to generate electricity, and the fuel needs to be transported to the facilities. Furthermore, spent fuels need to be replaced from the temporary storage sites to the final disposal sites. Spent fuel must be stored for one million years, during which fossil fuels are used for cooling the spaces. Throughout each process above, an enormous amount of energy is consumed and a huge amount of CO_2 is emitted.

The promoters of nuclear energy are well aware of this fact. The Japan Advertising Review Organization (JARO) judged in 2008 that *"(...) advertising nuclear energy as 'a clean way of producing electricity' as it doesn't emit CO_2 by generating electricity"* was inadequate. Lately, the expression has been changed to *"Nuclear energy doesn't emit CO_2 at the point that electricity is generated"*. However, it turns out that this statement is also wrong.

Kazuhide Sueda, of a trade union branch of the Environment, Agriculture, Forestry and Fisheries, member of the Osaka municipality, asked for all the following data accumulated and published from 2006 to 2008 to be disclosed: "system of estimates, report, and publication of greenhouse effect gas emissions" based on the global warming countermeasures promotion regulation. He analysed the disclosed data, and showed numerical facts that nuclear power plants emit CO_2 when generating electricity.

Take a look at Table 8. It's a table of the numeric result Mr Sueda has presented. The quantity of CO_2 emission when electricity is generated is indicated in the column "before distribution to power plants etc."

With boiling-water reactors as an exception, the cause of CO_2 emission is due to the use of supplementary boilers as their main system, which constantly burn crude petroleum as fuel. In the same article, we can find "the volume of greenhouse effect gas emission from each nuclear fuel facility" (Table 9, page 122). According to the table, even the facility called "Fugen", which ceased its operation in 2003 and which is currently in a decommissioning phase, still emits CO_2. The total quantity of CO_2 emission in 2008 from the two tables makes 116 tonnes. This amount is almost equivalent to the annual total quantity of CO_2 emission of all the operations in the Yamanashi prefecture (the third least CO_2 emission prefecture among 47 in Japan).

(unit: t/year)

Nuclear Power Plant	Reactor type (P: B pressurised water: boiling water)	FY2006		FY2007		FY2008	
		CO_2 related to energy	Power plants etc., before distribution	CO_2 related to energy	Power plants etc., before distribution	CO_2 related to energy	Power plants etc., before distribution
Tomari	P	-	-	-	-	27,800	0
Higashidori	B	19,100	6,430	19,300	7,030	24,300	5,500
Onagawa	B	71,700	6,210	8,380	8,380	4,770	4,770
Tokai Daini	B	16,300	0	11,500	7,030	16,100	6,750
Fukushima Daiichi	B	65,700	65,500	59,600	59,400	42,800	42,600
Kashiwazaki-Kariwa	B	8,780	8,780	4,980	2,740	241,000	8,470
Hamaoka	B	77,300	13,600	12,900	12,900	11,500	11,500
Shiga	B	60,199	8,548	109,000	8,300	70,900	9,220
Tsuruga	P, B	30,900	0	29,203	0	50,840	0
Shimane	B	25,200	9,920	16,500	10,700	32,200	10,200
Genkai	P	8,654	0	6,742	0	6,068	0
Sendai	P	-	-	6,100	0	5,534	0
Total		383,833	118,988	284,205	116,480	533,812	99,010

Source: from the graphic *The cost of nuclear plants,* Kenichi Ojima (Iwanami Shinsho)

Table 8: volume de CO_2 emitted by nuclear power plants

(unit: t/year)

	FY2006	FY2007	FY2008
Total of 3 enterprises with 4 fuel processing plants	29,276	27,813	27,992
Japan Nuclear Fuel (JNFL) - enrichment and landfill plants	26,400	19,500	17,700
Japan Nuclear Fuel (JNFL) - reprocessing plants	227,000	201,000	211,000
Japan Atomic Energy Agency (JAEA) - Monju	26,797	40,562	48,780
Japan Atomic Energy Agency (JAEA) - Fugen	16,831	17,245	19,390
Japan Atomic Energy Agency (JAEA) - total of other laboratories	275,440	145,560	301,720
Total	601,744	451,680	626,582

Source: Both tables 8 and 9 are based on the Kazuhide Sueda Article "The truth about the CO_2 emissions of nuclear power plants - the lie of "No CO_2 " (group 13, 33rd series of Circle Aichi Jichi)

Table 9: Volume of greenhouse gases emitted by the nuclear fuel facility

Discovering the new merit of thermal energy

Despite everything, if we compare the operation of electricity, it is true that thermal power energy emits more CO_2 than nuclear energy. Lately, however, thermal energy has made significant progress in implementing environmental countermeasures. For example, the Isoko thermal power plant in Yokohama has increased its gross thermal efficiency to about 45%, which is the highest by global standards. In doing so, CO_2 emission has been reduced by 17% kilowatt/hour. Conventional amounts of other toxic substances emitted by combustible carbon are successfully eliminated by: SOx (sulphur oxide), more than 95%; NOx (nitrogen oxide), more than 87.5%; and particulate matter, more than 99.9%. The ash after combustion is almost 100% recycled as raw material for cement. Carbon is the largest energy source originally existing on earth. Figure 18 shows the amount of solar energy we receive on earth in a year as well as the amount of underground natural energy resources. Solar energy is infinite while natural energy resources run out with consumption. These energy resources are called non-renewable energy. As you can see at a first glance at Figure 18, the largest square among the non-renewable energy is carbon.

The external square is the estimated quantity of the substances that exist on the earth (ultimate reserves). The black square in the interior represents the extractable quantity within the limits of current technology that can also be profitable (confirmed reserves). First, the small black square on the upper right indicates the total energy the world consumes in a year, which is 0.4. Thcn, thc confirmcd rcscrvc of carbon is 22.3, which will allow us to use nearly 56 years of electricity. If technology advances resulting in us reaching ultimate reserves, carbon alone will last for 775 years. Although their confirmed reserves are not shown in the figure, there are also resources such as oil shale and tar sand, which have been frequently used in the USA and Canada. As the exploration in this field has been quite fast, the squares in the Figure should be described as much bigger today.

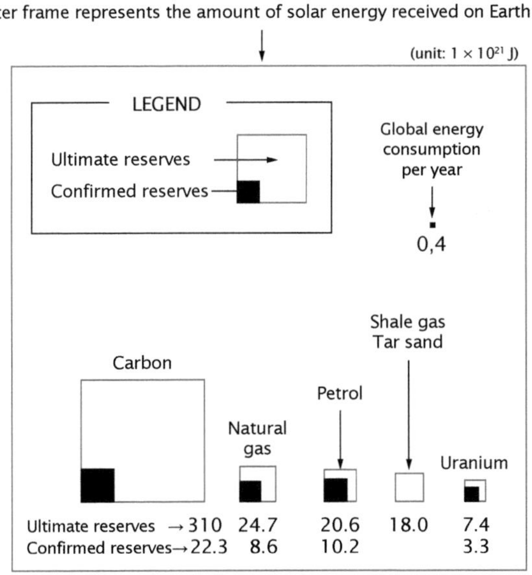

Source: Based on the data from "The energy demand by 2030, "Commission on the energy resources of the Agency for Natural Resources and Energy, and documents by the Tomabechi Committee, 4th meeting on ITER (26.05.1997)

Figure 18: Reserves of non-renewable energy resources

By the way, uranium on the extreme right side, fission uranium (235), which is used as fuel for nuclear energy, is only a fraction of that of petrol, the resource we believe is at risk of a shortage. Compared to carbon it is only a few per cent, which indicates that uranium is very scarce. The promoters of nuclear energy had to come up with a strategy to multiply the resource for nuclear energy, given the condition of a poor quantity of uranium. When uranium combusts, a new nuclide called plutonium, which does not exist naturally, is produced. The plutonium is burnt in a special facility called a high-speed breeder reactor. Then plutonium is extracted from the ember of fuel to burn it again. They imagined that this continuous cycle would increase the nuclear resource by up to 60 times more than the fission of uranium. However, the high-speed breeder reactor "Monju" has not yet started working due to numerous accidents. The facilities to re-process spent fuels are not even in operation as the technology has not been established yet. The dream had collapsed a long time ago, but even today, a huge amount of the budget (= our tax) is invested into this project.

Furthermore, the Government entrusted England and France with the re-processing of spent fuel to prepare Monju for the time of its operation. The total amount of plutonium, which has been preserved in Japan and overseas, was 44 tonnes in 2013[12]. This amount is equivalent to the capability of producing 4,000 atomic bombs of Nagasaki. Foreign countries criticised Japan for possessing such a high amount of useless plutonium as a non-nuclear power state. Reluctantly, Japan started the project called Pluthermal, using fuel called MOX (a mix of plutonium and uranium) burnt in normal nuclear reactors. This project is dangerous and economically non-profitable.

As for carbon, not only is there a sufficient amount, it can be found in many places on earth, which will allow stable distribution at a low cost. At the Isoko thermal power plant, which was mentioned previously, about 4,000 people a year from Japan and overseas have come to inspect

12 About 47 tonnes in 2014

it, including members of electricity companies and Government officials. Carbon and thermal energy are now re-acknowledged, and further investigation for energy efficiency and reducing CO_2 is in progress. When we look at the temperature rise and increase of CO_2 on the earth, it is true that the concentration of CO_2 in the atmosphere has been increasing. In my opinion, the cause of global warming, which has been discussed in our times, may not only relate to CO_2. It is not only me who thinks there are multiple causes. CO_2 is viewed as something bad, but it is an essential substance for human beings and our environment. Fission products, on the other hand, are toxic substances in all respects. The Fukushima Daiichi Nuclear Power Plant proves it.

Sooner or later, nuclear power plants come to an end with decommissioning

Before the second critical accident

The people who live around a nuclear power plant protest, "*If the nuclear power plant is stopped, we won't be able to make our livings.*" Well, this is exactly the objective of the "Nuclear Mafias".

The communities who accepted the construction of a nuclear power plant tend to make comments such as: *"it was a national project, therefore, the Government should look after us"* and *"we want another industry for a replacement, otherwise, we will lose everything."* People's reaction to blame it on the State is quite normal, and I have nothing against that. As we can see, however, neither the State nor promoters of nuclear energy will take any responsibility. They are not even trying.

"The promoters of nuclear energy" include the people who accepted the construction of a nuclear power plant in their hometown.

For more than 40 years, I have been repeating that, *"nuclear power plants are dangerous. A critical accident will happen one day, and when it happens, the damage will be catastrophic."* I was involved with the protests against accepting the construction of nuclear plants. Unfortunately, the anti-nuclear civil movements did not grow anywhere near the size of the gigantic power of the nuclear mafias. In fact, the habitants of the sites knew the danger of nuclear power plants. Despite their knowledge, they were blinded by PR campaigns such as *"nuclear energy is an ecological energy that doesn't harm the environment"* and *"we will run out of electricity without nuclear energy"* together with the large amount of subsidies laid out before their eyes.

On the other hand, there were communities who refused to accept the construction of nuclear plants and chose to live independently without receiving the subsidies. The numbers of those municipalities are three times more than those who accepted the constructions. Those people did not believe in sugar-coated words such as nuclear energy is absolutely safe. They put their questions to the electric companies and politicians, *"Why don't you construct a nuclear plant in Tokyo or Osaka, if you say that nuclear energy is absolutely safe?"* The answers were not clear enough for them to understand them.

The habitants of the communities who accepted the construction of nuclear plants should have been aware of this grey-zone. They had to make a bitter choice in order to get out from difficult financial struggles and rebuild their communities economically. It should not be explained as "they believed the safety myth about nuclear energy", but rather "they chose to believe the safety myth about nuclear energy".

① Hokkaido, Wakkanai
② Hokkaido, Hamamasu (now known as Ishikari)
③ Hokkaido, Kitahiyama (now known as Setana)
④ Hokkaido, Taisei (now known as Setana)
⑤ Hokkaido, Matsumae
⑥ Aomori, Shiura (now known as Goshogawara)
⑦ Aomori, Higashidori, Gamanosawa et Noushi
⑧ Aomori, Kamikita
⑨ Iwate, Kuji, Samuraihama, Honnami
⑩ Iwate, Tanohata
⑪ Iwate, Taro (now known as Miyako)
⑫ Akita, Noshiro, Asanai
⑬ Akita, Yurihonjo, Iwaki-kameda, Kameda, Tsuruoka
⑭ Fukushima, Namie et Odaka (now known as Minami-soma)
⑮ Niigata, Maki (now known as Niigata)
⑯ Ishikawa, Suzu
⑰ Fukui, Kawanishi (now known as Fukui), Sanrihama
⑱ Fukui, Obama
⑲ Mie, Kisei (now known as Taiki) et Nanto (now known as Minami-ise), Ashihama
⑳ Mie, Kiinagashima (now known as Kihoku), Jonohama
㉑ Mie, Miyama (now known as Kihoku), Ojirohama
㉒ Mie, Kumano, Ichiura
㉓ Kyoto, Maizuru
㉔ Kyoto, Miyazu
㉕ Kyoto, Kumihama (now known as Kyotango)
㉖ Hyogo, Kasumi (now known as Kami)
㉗ Hyogo, Hamasaka (now known as Shinonsen)
㉘ Hyogo, Mitsu (now known as Tatsuno)
㉙ Wakayama, Nachikatsuura, Taiji
㉚ Wakayama, Koza (now known as Kushimoto)
㉛ Wakayama, Hikigawa (now known as Shirahama)
㉜ Wakayama, Hidaka, Ao
㉝ Wakayama, Hidaka, Oura
㉞ Tottori, Aoya (now known as Tottori)
㉟ Shimane, Gotsu, Kuromatsu
㊱ Shimane, Masuda, Takatsu
㊲ Okayama, Hinase (now known as Bizen), Kakuijima
㊳ Yamaguchi, Tamagawa (now known as Hagi)
㊴ Yamaguchi, Hagi
㊵ Yamaguchi, Hohoku (now known as Shimonoseki)
㊶ Tokushima, Anan
㊷ Tokushima, Hiwasa (now known as Minami)
㊸ Tokushima, Kainan (now known as Kaiyo)
㊹ Ehime, Tsushima (now known as Uwajima)
㊺ Kochi, Kubokawa (now known as Shimanto)
㊻ Kochi, Saga (now known as Kuroshio)
㊼ Fukuoka, Shima (now known as Itoshima)
㊽ Fukuoka, Shimakoganemaru (now known as Itoshima)
㊾ Kumamoto, Amakusa
㊿ Oita, Kamae (now known as Saiki)
(51) Miyazaki, Sadowara (now known as Miyazaki)
(52) Miyazaki, Kushima
(53) Kagoshima, Uchinoura (now known as Kimotsuki)

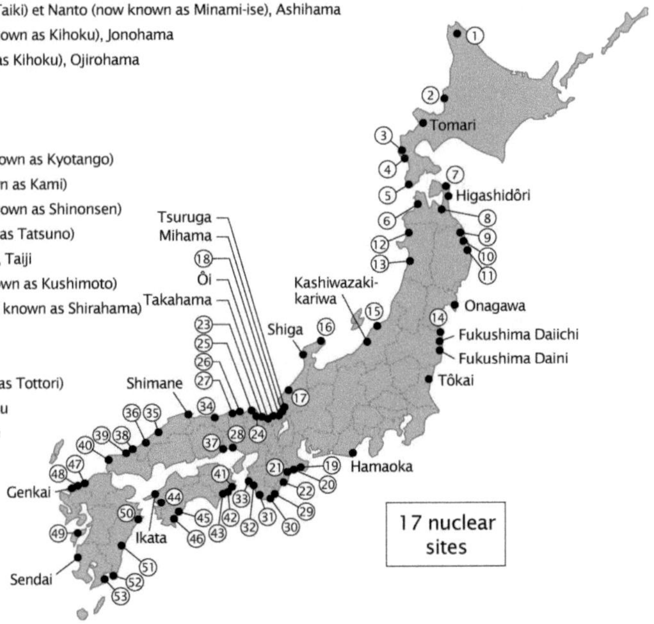

Source: Based on the following books: booklet of Japan Catholic Council for Justice and Peace "Nuclear energy is not a weapon against global warming!" "Those who refuse the construction of nuclear power plants - Iwai Island to the future, Shin Yamaaki (Iwanami Shinsho)," "The communities who have said no to the nuclear and anti-nuclear movement" Yuko Hirabayashi (Laboratory Journal Ohara on social issues, Hosei University, No. 661, November 2013)

Figure 19: The communities who refused the construction of nuclear power plants are three times more than those who accepted them

However, I don't think those people were simply deceived.

Even if one is not directly involved in promoting nuclear energy, the whole nation, except children, is partly responsible. Not to mention the promoters, who should be held accountable for even greater responsibility. Each situation is different, but everyone should take his or her fair share of the blame.

Nuclear power plants, even if they are not stopped immediately, should be decommissioned sooner or later. You may think the old reactors, which come to the end of their useful life at the sites with several reactors, can simply be replaced with new reactors so the life of the plant can be extended indefinitely. The habitants in Fukushima might have believed that too. However, we should realise that nuclear power plants are machines. Obviously, machines can break or have problems. With Japan as a seismic state, and as an island that is particularly vulnerable to all kinds of natural disasters, it is impossible to believe such a myth. Even if one reactor successfully completes its useful life during our generation, there is no guarantee that it will last for the next generation. On the contrary, if nothing happens in our time, the probability of risk will only increase in the future. We cannot pass something so dangerous onto our future. I have been advocating this message for a long time. Unfortunately, the message hasn't reached the people whom I most want it to reach.

One day, peaceful, everyday life is suddenly destroyed. It must be a tremendous shock. However, we can only appreciate its value after it is gone. That must be the case for the plaintiffs of the action group against the Fukushima Nuclear Power Plant.

The nuclear power plant, which was claimed to be absolutely safe, suddenly caused an accident. Several hundreds of thousands of people were forced to leave their hometowns, deprived of their lives and a place to go back to.

Several thousands of deaths were related to nuclear accidents. The vast lands spread around the site, which have become no-entry zones, should be designated "radiation controlled areas" by Japanese law.

The state of Japan abandoned several million people in the contaminated areas due to the "emergency situation". Only a small percentage of the population were capable of escaping from there. Those who couldn't leave, including babies, have no choice but to continue living in the contaminated areas.

I cannot begin to imagine the anxiety, stress and horror, which engulf the plaintiffs in this article. I hope their voices will be heard and people will come together to face the problems.

Extracted article: "Reading "Where is the accountability?" by the action group against the Fukushima Nuclear Power Plant (Friday) - Day citizens judge power crime, Hideaki Koide", Tosho Shinbun (no. 3134, 16 November 2013)

Change required for a new way of living

It is a big challenge to rebuild the lives of habitants who depend on the nuclear power plants. In the end, they have to stand on their own two feet by themselves. The measures should not be offered by the State, but taken seriously by the habitants.

Accepting the difficulties and agony in working and living, they should look for a way to survive.

Yet, a possibility does exist: the decommissioning operation produces another 40 years of work. If we look worldwide, 538 nuclear reactors with the power of more than 10 MW, among which 149 of them have already been stopped for decommissioning. The list of decommissioning reactors includes the Fukushima Daiichi Nuclear Power Plant (1-4 reactors), the Hamaoka Nuclear Power Plant (1 to 2 reactors), the Tokai Nuclear Power plant, and Fugen. Later, reactors 5 and 6 of the Fukushima Daiichi Nuclear Power Plant also joined the list.

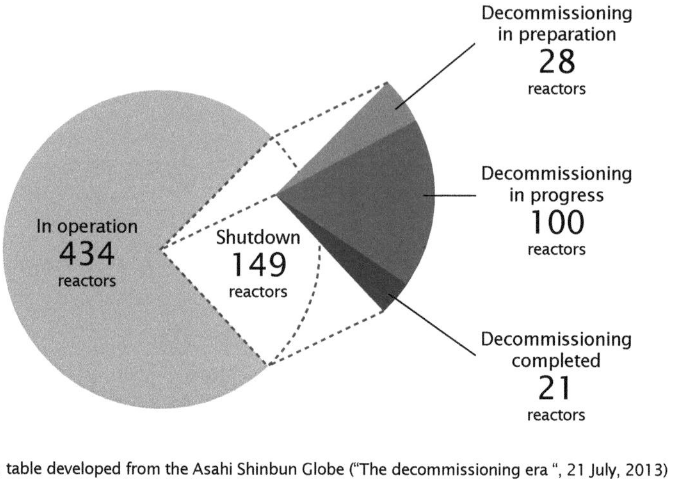

Figure 20: Nuclear power plants worldwide coming to an end (data from the Radwaste and Decommissioning Centre (RANDEC), as at 27 June 2013)

The world has already entered into the era of decommissioning, and there are many specialised growing businesses for efficient and safe decommissioning operations. The specialists working at the sites are the engineers who used to operate the nuclear reactors. I came across a special feature in a newspaper that such kind of entrepreneurship exists in the Niigata prefecture, Japan. ("The era of decommissioning" in "Asahi Shinbun Globe" 21 July 2013). In Japan, among the 48 reactors left, 16 of them have been operating 30 years and 3 of them are 40 years old. (As at February 2014).

It looks like the decommissioning era will arrive before the nuclear politics will be resolved. Here, what concerns me is that Japan is not educating new specialists enough on nuclear energy for the future.

Date of entry into operation	Number of years of operation (as at February 2014)	Reactor
March 1970	43	Tsuruga No.1
November 1970		Mihama No. 1
July 1972	41	Mihama No. 2
March 1974	39	Shimane No. 1
November 1974		Takahama No.1
October 1975	38	Genkai No.1
November 1975		Takahama No. 2
December 1976	37	Mihama No.3
September 1977	36	Ikata No.1
November 1978	35	Tokai Daini
March 1979	34	Oi No. No.1
December 1979		Oi No.2
March 1981	32	Genkai No.2
March 1982	31	Ikata No.2
April 1982		Fukushima Daini No.1
February 1984	30	Fukushima Daini No. 2
June 1984	29	Onagawa No.1
July 1984		Sendai No.1
January 1985		Takahama No. 3
June 1985	28	Fukushima Daini No. 3
June 1985		Takahama No.4
September 1985		Kashiwazaki-kariwa No.1
November 1985		Sendai No. 2

Source: Based on the data in the document "The nuclear power plants in operation and under construction in Japan," Japan Atomic Industrial Forum (JAIF)

Table 10: List of nuclear power plants starting with the oldest

As nuclear power plants prospered all over the world, their problems and dangers too were revealed. Little by little, we start to realise that nuclear energy has no future. Since about 20 years ago, the number of the students who chose nuclear science as a major has decreased drastically. All the facilities in seven ancient imperial universities removed the word "atomic" from their subjects. Only a few limited universities offer the course. The Ministry of Education and Science offers grants but universities cannot open the facilities as it is hard to attract enough students. Nevertheless, we will definitely need specialists for the future for the decommissioning operations. In addition, something must be done about the high level of radioactive disposals, the quantity of 1.2 million Hiroshima bombs. However, it is difficult to ask the younger generation to sacrifice their careers to take care of the waste produced by previous generation who dreamed and failed in a fantasy for nuclear energy.

Peace cannot be built on the logic of power

The Japanese Government's ambitions and America's aim

Today, the world is governed by the United Nations. If we translate this term literally, it is the nations united by allies: the allies who were united to battle against Germany, Italy, and Japan during the previous world war. In the United Nations Charter, there is something called the article against hostile countries. In this article, it states that, Germany, Italy and Japan can be treated in any way as they are hostile countries. This article still exists today. More than 50 nations have participated in the united allies. The reason why only the USA, Russia, the UK, France and China could become permanent members of the Security Council was because they have atomic forces. Possessing atomic forces is a crucial factor in dominating today's world. It is normal to think that possessing nuclear weapons is indispensable for being powerful in this world.

Those permanent members of the Security Council possess nuclear weapons, but don't allow other nations to possess them. The Nuclear Non-Proliferation Treaty is a system for them to monopolise nuclear weapons.

Then, thanks to the international "Nuclear Mafias" called IAEA (International Atomic Energy Agency); they created a system to check upon other nations. As you can imagine, other nations were not happy about it. It is natural that other nations strongly wish to possess nuclear weapons and become a permanent member of the Security Council that dominates the world.

The leaders of many countries are obsessed by this idea of *"nuclear weapons are indispensable to dominate the world"*, and wish to possess nuclear weapons.

The Japanese Government LDP is also obsessed by this idea. In fact, Japan has signed the Japan-United States atomic agreement (for collaboration between Japan and the United States in terms of the peaceful use of nuclear energy). In the name of peaceful use, Japan has obtained the technologies for enriching uranium, nuclear reactors and reprocessing, which enabled Japan to produce nuclear weapons. However, Japan is the only nuclear state which is declared a non-nuclear state. The reason why the United States agreed to such conditions was originally because this atomic agreement was linked to the Japan-United States Security Treaty and Force Agreement, which indicates that Japan would become a dependency of the United States. As for the United States, the agreement was willing to give certain freedom to its dependency. Having accepted the agreement, the United States hesitated in allowing Japan to possess the capability to use the third technology, reprocessing, because plutonium can be produced by reprocessing, Plutonium is the ingredient to make atomic weapons. With this technology, atomic weapons can be constructed at any time. As for Japan, that was precisely the reason for implementing the reprocessing system. In 1977, at a time the United States appealed to the world to stop the reprocessing of spent-fuels as a step for nuclear non-proliferation, there were some delicate conflicts between the two countries. Ultimately, Japan won the negotiation to mix uranium with plutonium, which was once separate. The United States tolerated the request by Japan. Finally, with the support of France, the Tokai Reprocessing Plant began to operate. The reason for the French support was that, at that time, the United States had abandoned commercial reprocessing to promote the movement for nuclear non-proliferation.

Other countries such as Korea and Iran criticised the United States, claiming the special treatment given to Japan was a double standard. Despite those criticisms, the United States didn't make Japan stop using nuclear energy, because they wanted to use Japan as a dependency. Why? To make money.

The United States started to withdraw from nuclear energy in 1974. They no longer have the production lines to construct nuclear reactors. However, they still possess the patent to make nuclear reactors. Moreover, as they have built so many facilities for uranium enrichment, they have plenty of surplus enriched uranium. Therefore, they want to force Japan to buy nuclear reactors with nuclear fuels on top. The more they sell, the more money they make. By keeping Japan inside of the circle of nuclear states, they can use production lines in Japan, selling nuclear energy to other Asian countries in order to make even more profit. Of course, they didn't forget to factor in restraining the threats from China. Japan, on the other hand, doesn't wish to be excluded from the framework as a dependency of a dominant state, the United States. The nuclear production companies can no longer abandon the business because they have already constructed the production lines, even though they won't be able to construct any more plants in Japan. Therefore, they want to sell them to overseas. In such a way, the complex mix of calculations of political and economic interests made two countries inseparable from the nuclear industry.

The term of the Japan-United States atomic agreement will expire in 2018. I think it should be abrogated. Japan must search for a true sense of independence from the United States.

The Japanese Constitution

Possessing atomic power is the power to dominate the world, that's what LDP members believe. However, peace cannot be built on the logic of power. We should all realise it by now. It is extremely difficult to maintain freedom and peace. It is even more difficult to establish freedom and peace.

In the preamble of the Japanese constitution, it says:

> *"[We, the Japanese people]...resolved that never again shall we be visited with the horrors of war through the action of Government, do proclaim that sovereign power resides with the people and do firmly establish this Constitution."*

In short, the constitution signified that it is citizens' determination to check on our Government to make sure it doesn't do anything stupid. Each individual has to be responsible, otherwise neither freedom nor peace can be established, and another war could happen due to our lack of responsibility. That is what the preamble states. It also says:

> *"...we have determined to preserve our security and existence, trusting in the justice and faith of the peace-loving peoples of the world."*

It is not a military force that guarantees our security. Japan is determined to preserve our own security by trusting in the justice and faith of people. This is extraordinary and not easy at all. The constitution states that citizens are responsible for something of such importance.

It also says:

> *"We recognize that all peoples of the world have the right to live in peace, free from fear and want."*

It's not only Japan. The constitution recognizes everyone in the world has the right to live in peace. Our pacifism is not limited to our own nation. It says that Japan renounces its arms not only for Japan's pacifism but also for the rest of the world. It says Japan believes in trusting in the justice and faith of every citizen in the world. It is a beautiful constitution.

Japan's nuclear technology is unreliable

Japan's interest in the exportation of atomic technology

Today, with Prime Minister Abe as a top salesman, Japan is accelerating to sell nuclear power plants to overseas. Since all nuclear power plants have been stopped with no promise for re-starting, and no hope for future construction in Japan, the Japanese nuclear industry is forced to find a means of survival focused towards exportation. At the same time, Japan has a deep enough understanding that nuclear energy is so dangerous that it cannot be accepted in its home country. It is such a pity that Japan, with the constitution that hopes for world peace, imposes nuclear business on foreign countries. It must be stopped.

In Eastern Europe and developing countries in particular, demands for nuclear power plants are increasing. By 2030, about 90 to 370 reactors are planned for new constructions. Why is there such a large need for nuclear power plants? It seems to me the objective is not purely to generate electricity. As explained before, they want to possess atomic weapons too.

In Japan, there is a distinction between the terms "nuclear" and "atomic". It seems that nuclear energy is for peace, and atomic power is for arms. In fact, these terms have the same meaning. Japan's objective to engage with the nuclear energy industry, although it is recognized as peaceful use, was also to obtain plutonium, which is the ingredient for atomic weapons.

Nuclear energy is the same as atomic energy. It is an atomic weapon itself therefore exporting nuclear technology is equivalent to exporting atomic technology. We should realise that Japan is involved in such horrifying acts.

Japan as an underdeveloped nuclear country

Sometimes, I find articles in newspapers worrying about Japan's top-level nuclear technology being drained to overseas. Many people seem to misunderstand, but Japanese nuclear technology is not necessarily high-level. Japanese people believe that Japan is advanced in the field of technology and science, but it's not true. As far as nuclear technology is concerned, Japan is a rather underdeveloped state.

In 1977, the Tokai Reprocessing Plant was constructed by France. As stated previously, at that time, Japan was dependent on France and England for reprocessing. By learning manufacturing and techniques from France and England at Tokai, Japan proudly announced that they would construct the Rokkasho-Village Reprocessing Plant only with domestic technology. As a result, the project failed, and Japan had to ask France to construct it. Embarrassed for the failure, Japan took charge of vitrifying high-level radioactive effluent, which is discharged at the reprocessing. However, it also failed, and now the process has been halted ever since.

In 1999, a critical accident occurred at the uranium conversion facility of Tokaimura, the Ibaragi prefecture. It's called the "Tokaimura JCO critical accident". When I first heard the news about this accident, I could not believe the critical accident had really happened. Because a critical accident is something that should never happen, although several accidents had happened in the world in the past, it was already eradicated at the beginning of the 1970s. It has become a common understanding among the nuclear field that such an accident would never happen again. That is to say, with natural uranium dispersed on the earth with thin density, an accident would never happen under such conditions. It had become clear what causes a critical accident: the point at which; the amount at which; and the way in which uranium should not be collected. Every country in the world took care of uranium with collective knowledge. On the contrary, at JCO's facility, the State approved the use of a settling tank. The workers who did not have knowledge about the critical accident put too much uranium liquid in the settling tank. That's how the accident happened. The cause of the accident was reported as the workers did not follow the manual. However, the promoters of nuclear energy kept saying to the workers that the situation would be safe even if something went wrong or if workers made mistakes, by using phrases like "fail safe" and "fool proof". It is true that the workers did not follow the manual, but the facility should have been constructed to avoid accidents from happening.

This incident proved that Japan is a country that cannot take even basic control over uranium.

Japan has constructed 58 nuclear reactors to date. Each one has a safety guarantee by LDP, but many accidents, large and small, have already occurred. Then in March 2011, to my greatest regret, a critical accident happened at the Fukushima Daiichi Power Plant. Since then, though more than four years have passed, there is not even an answer to control the situation. Many workers have been exposed to radiation, and it is not possible to block radioactivity from going outside. This is the reality. It is the clear proof that Japanese nuclear technology is not at all reliable.

Exporting nuclear technology overseas entails a great financial risk to the enterprises. For example, at Mitsubishi Heavy Industries, coolant water, including radioactive substances, leaked from the pipes of the vapour generator installed at San Onofre nuclear power plant in California, which led to the decommissioning of the plant.

In October 2013, the buyer, Southern California Edison Company, claimed 4 billion dollars (about 4 hundred 1 billion yen) for the damage, including the cost for decommissioning. This affair was taken to court. The cost charged was three times the ordinary income of Mitsubishi in financial year 2012.

Compensation for damage is not the only risk. In the same article, the first order received was only tentative. After agreeing on the general framework, environmental conditions such as geological features and economic conditions need to be investigated for several years. If there is a slight problem, for example, an active faulting at the planned construction site, the contract may be cancelled. If the company doesn't have solid strength to resist against obstacles, the project cannot be continued. The Prime Minister Abe, with a big smile on his face, is busy with his sales promotion. By selling nuclear technology, the more that is sold, the more risks of nuclear accidents on the earth increases. Also, it is an equivalent act to spreading atomic weapons and threatening world peace. On top of that, the risks the makers of nuclear power plants have to bear are enormous.

Nuclear energy as a mark of discrimination

Those who don't receive the benefit of electricity have to make sacrifices

The former Prime Minister Junichiro Koizumi's "*zero nuclear energy immediately*", is rippling through the Japanese society. His argument *"Because there is no vision for permanent disposal sites, we should quit nuclear energy"* is clearly logical. LDP and/or their financial supporters, people in the electricity industry criticise Koizumi's comment as irresponsible. In my opinion, postponing the issues of final disposal sites and looking for instant profit is much more irresponsible than Koizumi's comment.

What do members of the Diet think about the issues of final disposal sites? In Shukan-Asahi, 20 December 2013, the results of a questionnaire to all members of the House Representative regarding the nuclear policy were published. According to the results, 110 out of 480 members responded and returned the questionnaire. Among them, 58 answered the question on "geological disposal of high-level radioactive disposal": 28 answered as "not possible in Japan", 21 answered as "mandatory to be constructed in Japan", and 9 answered "there are other solutions besides constructing it in Japan". More than half of the members who answered the questionnaire agreed to dispose of radioactive waste somewhere but Japan. Where should that be? Not under the sea, not above the sky... then, it will be on the lands of other countries. Actually, at the sales front of nuclear technology to overseas, Japan offers bonus points, which means the client country can return the nuclear waste to Japan. This is a sales strategy in order to win the competition against Russia and Korea. Japan cannot even find a place to dispose of its own radioactive waste. It turns out that Japan is planning to return the radioactive waste to where uranium is extracted.

Much to my surprise, the sites where uranium is found tend to always be in indigenous people's territories, such as in the United States, Australia, India, and Mongolia. Those people have received very little merit from electricity. They provided manual labour and were exposed to radiation, and are now at risk of irradiation by returned uranium waste. History has always been like this. Once, Japan put forward Mongolia as a final disposal site, but Mongolia refused.

In addition to the result of the questionnaire above, Koji Sakamoto, deputy of LDP, chosen from the evacuated area of Fukushima commented as follows:

"I think Japan should construct a final disposal site like "Onkalo" in our state. In that case, it should be linked to the delocalisation of administrative functions of the capital. Unfortunately, my idea is not having much impact.

The final disposal site should be constructed on the most stable ground. The most stable ground is also most suitable for capital administrative func-

tions. The radioactive waste should be disposed of directly underneath the new parliament building with all parliament members working there.

To prove the safety to citizens, Diet members should demonstrate their commitment."

Extracted from the article "Constructing Onkalo under the Parliament Building" (20 January 2013). Shukan Asahi

I completely agree with his idea.

Not to make another mistake

Nuclear Mafia

The Nuclear village is a collective criminal group

As you may have already noticed, when I talk about the promoters of the nuclear industry as a gigantic power structure, I am using an expression "Nuclear Mafias" instead of Nuclear Village". Calling them a "Nuclear Village" is inadequate. That is to say, the term "village" is a community system composed of various kinds of people, and the organisation of a nuclear village has that aspect. However, it's more like a criminal collective than a community.

In Japan, it was decided that the State would take charge of the nuclear industry. Setting up the Electric Utility Industry Law, the State created a system that allows automatic profit to be generated by developing more nuclear energy. Then the electric companies were drawn into the system. A gigantic industry, which represents Japan such as Hitachi, Mitsubishi, Toshiba and general construction and other companies around them, participated in profiting from nuclear energy.

Mass media, academics, and courts, even unions are associated with and promoted nuclear industry. And each one of them knew that nuclear power plants pose an enormous danger. As proof of this, they have never chosen to build a nuclear power plant in a big city. Instead, they have imposed the plants on rural villages and towns that have financial difficulties. Nonetheless, no one tried to stop the outrageous development of the nuclear industry. In the end, the irreparable accident happened at the Fukushima Daiichi Nuclear Power Plant.

The whole movement around nuclear power plants is very similar to the situation during the previous war. During the Second World War, there were people, even in the military, who knew that the war they were involved in had no hope of victory. But no one was able to express it. Most of the people were engulfed by the huge wave.

Similarly, the "nuclear mafias" - shutting the mouths of powerless people by money and power, and proudly flying the flag for nuclear energy - are nothing but criminal collectives. They deserve to be called "nuclear mafias".

Many lives were lost because of the accident at Fukushima

"No matter how much you criticise, reality shows that no one died from the nuclear accident", counter-argue some promoters of the nuclear industry. I was shocked to hear that. Many people have already died as a result of the accident.

For example, there is a big hospital called Futaba in the Okuma Village, where the Fukushima Daiichi Nuclear Power Plant was located. The hospital was located four kilometres away from the plant. At the hospital and related institutions, there were 440 patients. That day, when the successive explosions happened, radioactivity dispersed, and the dosimeter level started to rise. Eventually, even police officers started to say they should evacuate. The hospital workers must have felt bad, as transferring patients entails great risks. However, they had no choice but to run to safety as a radioactive cloud was approaching. With unimaginable physical effort, they transferred patients by putting them on stretchers and then moved them to the cars. During the transportation of the patients, some passed away. The death toll reached 45 people. (Mainichi Shinbun 26 April 2011).

Incidents like this have not only happened at the hospitals. In houses, people who could not move by themselves died without being saved. Among the people who evacuated, many elderly people lost their lives at the shelters.

There was a man who owned a dairy farm near the plant. He hung himself by leaving the message on the cowshed wall, *"if only there was no nuclear power plant..."*. Another farmer hung himself in the middle of his cabbage farm by looking at his contaminated cabbages. It's not just humans who have died. Animals such as cows, horses and pigs that were like families to the dairy farmers, as well as pets such as dogs and cats have also died. They were left behind and lost their lives. Some of the owners of the animals couldn't bear leaving them tied up, so they released them. If the freed animals came out of the radiation controlled areas, the guards sent by the Government would shoot them to prevent the radioactivity from spreading.

An unforgivable crime

What is most unforgivable is, even after all this has happened, Japan is still strongly promoting the nuclear industry. The Fukushima Daiichi Nuclear Plant, which was certified as "safe" by the State, caused an unacceptable accident. Not only is it far from under control, but even the cause of the accident hasn't been determined and the responsibility hasn't been clarified. Despite such a situation, the Prime Minister Abe, who is leading the LDP Government, is promoting the restart of the nuclear power plant after *"confirming its level of safety"*. Further, he declares that Japan will construct a new nuclear power plant, and Japan will import nuclear plants using the "No.1" nuclear technology in the world. This is beyond belief. Besides, these new agendas were approved immediately after the Fukushima accident by the bureaucrats in the Ministry of Economy and Industry.

In September 2013, the Prime Minister Abe stood in front of the world to bid for the Olympic Games in Tokyo, and successfully convinced the world by promising that the contaminated water problem was *"under control"*. What a joke! Before bidding for the Olympic Games, Abe should have saved the people he had let down in Fukushima.

Incidentally, Asahi Shinbun and the Fukushima University during the end of October and the middle of November 2013 conducted the fourth questionnaire given to the evacuated habitants from the Fukushima Nuclear accident. In the responses from the habitants on the question, *"Do the future Olympic Games in Tokyo have any impact on the reconstruction of Fukushima?"*, 34% answered *"bad impact"* which was more than double the 14% who answered *"good impact"*.

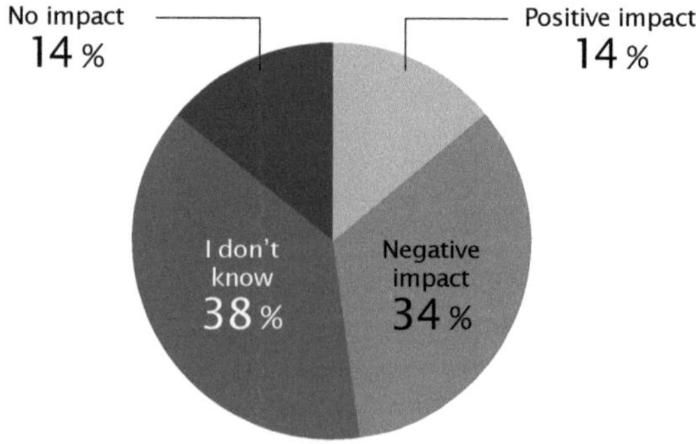

Source: Based on the data in the document "The nuclear power plants in operation and under construction in Japan," Japan Atomic Industrial Forum (JAIF)

Figure 21: Do you think that hosting the Olympic Games in Tokyo will have an impact on the reconstruction of Fukushima?

TEPCO must be closed down

Then the largest and most direct responsibility, needless to say, falls on TEPCO. This electricity company in Tokyo must be brought down by any means. Unfortunately, as the banks continue to provide finance in order to save TEPCO, many people are trying to ensure TEPCO survives. For example, neither the chairman nor the president has been condemned. Other board members, executives, and employees remain in high-salaried positions. Their salaries were cut by 20-30% at the time electricity costs increased, but that is not sufficient compared to the average salary of Japanese people, and furthermore, considering the hardship of the people in the affected areas.

Although those injustices were overlooked, the world is now scrutinising Japan. On 2 December 2013, KLP, the life insurance company in Norway, announced that it had sold its shares in TEPCO. The reason was of course "Fukushima". *"Although three years have passed, the situation is still not under control. The risk of further radioactive contamination remains an issue."* commented the investment related employee of the KLP. (Reuters, 2 December 2013). On the same day, a following article was published: *"1,700 TEPCO workers have requested voluntary retirement".* (Kyodo Tsushin, 2 December 2013). The continuous salary cuts and anxiety about an uncertain future were the main reasons. They have been paid high salaries for a long time. Once the accident happened, they just ran away from the responsibility (voluntary retirement). I find such conduct totally outrageous.

Erasing memories and concealing information

"Fukushima" as if it never happened

These days, we hear less and less news about Fukushima. The reality of how victims are still suffering is rarely reported. In order to *"re-start nuclear power plants"*, *"construct a new plant"* and *"export the nuclear technology to overseas"*, it is necessary for the Nuclear Mafias to make people forget about Fukushima. In short, they are trying to paint over the reality as if the Fukushima accident never happened. Not surprisingly, mass media is following this trend.

The above-mentioned questionnaire to the evacuated habitants from the Fukushima Nuclear accident includes the question regarding the fading memory of the accident. The results of the question show that more than 80% of the evacuated habitants feel that they have been forgotten by the rest of the Japanese nation (Figure 22).

Japanese people have a tendency to think *"out of sight, out of mind"*, so it may not just be because of the effect of the Nuclear Mafias. But although the Nuclear Mafias are trying to make us forget, we should be committed to not forgetting about Fukushima.

Nuclear energy: the energy for our bright future
Nuclear energy: local dynamism and future prosperity
Nuclear energy: enriching society and developing towns
Nuclear energy: prosperity in life through thorough understanding

These are the slogans posted on the arch-shaped signboards, still standing alone in the ghost town, Futaba, where the entire population was forced to evacuate. The slogans were invented by the school children at the time, which was 27 years before the accident. The photo of Katsutaka Idogawa, the Mayor of Futaba town, in white radiation protection gear, standing under the signboard "Nuclear energy: the energy for our bright future" was published on the front cover of a magazine. In the comment below the photo, he says, *"it is ironic; far from a bright future, the entire habitants had their lives taken away. The town is colonised by radioactivity."* And *"We miss Futaba. This signboard should be preserved as a negative legacy of mankind. We don't want nuclear power plants any more"*. (Shukan Gendai, 30 November 2013).

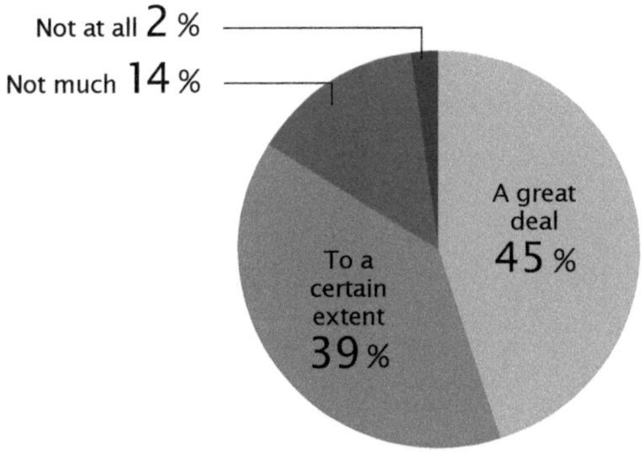

Source: Taken from the Asahi Shinbun ("The earthquake northeast of Japan, 1000 days later", December 4, 2013)

Figure 22: Do you think the memory of the accident at Fukushima is fading away and has been forgotten by the nation?

Approval of the Protection of Specially Designated Secrets Act

In Japan, there is a law called the Atomic Energy Basic Act, which was enacted in 1955. The fundamental policy of this law states that, *"the utilization of nuclear energy shall be limited to peaceful purposes, shall be aimed at ensuring safety, and shall be performed independently under democratic administration, and the results obtained shall be made public so as to actively contribute to international cooperation."*

Despite the Statement *"limited to peaceful purposes"*, the substantial purpose is to have the potential to make atomic weapons. As stated previously, Japan has accumulated a large amount of plutonium to date. Just in case you might think the information above is only my delusion, here is the proof:

> *"For the time being we will maintain the policy of not possessing nuclear weapons. However, regardless of joining the NPT or not, we will keep the economic and technical potential for the production of nuclear weapons, while seeing to it that Japan will not be interfered with in this regard. Furthermore, citizens will be enlightened to understand that, the general nuclear policies will be decided upon based on the calculation of international politics as well as economic gains and losses. We hope to avoid unnecessary internal panic in the unlikely event of strategic use of nuclear weapons in the future."*

This paragraph has been taken from "The Outline of our Foreign Diplomatic Policies" in 1969. Forty-five years ago, Japan had already established such a solid strategy and has been promoting nuclear energy ever since.

Also, one of the executives of the Ministry of Foreign Affairs commented in the newspaper, in response to the Japanese diplomatic force, that, in his personal (not public) opinion, *"it's better to leave the option open to choose atomic arms".* Adding *"We have the capacity to possess nuclear weapons, although strategically, we choose not to possess them following the current policy". Therefore, "it is necessary to stock plutonium as well as to develop technology to transform rockets into missiles".* (Asahi Shinbun, 19 January 1992).

As an individual who is involved in the nuclear domain, I consider that nuclear energy is completely dangerous. Under such risky intention of the Japanese Government, it becomes even more important to keep an eye on them.

I am not the only one opposed to nuclear energy. About 70% of the citizens are against the use of nuclear energy. Then, the Protection of Specially Designated Secrets Act, despite about an 80% opposition from citizens, was steamrollered through the plenary session of the House of Representatives. At the public hearing held in Fukushima the day before, all the people who stated opinions, including the supporters of LDP, were against this law. Many international organisations and groups, including the specialists of human rights from the United Nations, have issued proclamations.

Both for nuclear energy and the Protection of Specially Designated Secrets Act, so many people were raising their voices, but the Government is not listening.

Now that the Protection of Specially Designated Secrets Act has been approved, I am concerned that the information regarding nuclear energy will be concealed. Specifically, in June 2012, the Atomic Energy Basic Act was revised. In the basic policy introduced previously, the following sentence was added:

> *"Concerning the objective of ensuring safety, stated in the previous article, it shall be performed under the established international standard, with the purpose of protecting the lives, health, and welfare of the population, as well as contributing to protecting the environment and national security."*

"National security" is, as we are aware, much like the U.S-Japan Security Treaty: military jargon. In short, it implies *"nuclear energy can be used for national security"*. In "The Outline of our Foreign Diplomatic Policies" in 1969, it states, *"the citizen will be enlightened to understand that, the general nuclear policies will be decided based on the calculation of international politics as well as economic gains and losses."* That means that whether Japan will possess nuclear weapons or not is a matter of political strategy, and it will be decided based on the calculation of gains and losses, which the Japanese people should be aware of. Today, I assume that the Government is just starting to exercise the law. The problem is, however, the words *"contributing to*

national security" enable the information to be treated as a specially designated secret. The information of nuclear energy has never been fully disclosed. With this new act, it will become even more difficult to know the truth.

"[We, the Japanese people]... resolved that never again shall we be visited with the horrors of war through the action of Government, do proclaim that sovereign power resides with the people and do firmly establish this Constitution."

It is citizens' responsibility to watch their Government's abuse of power. Under the national liberalism and non-violation of human rights, it is the norm of the Government to protect the privacy of individuals, whereas the State is keeping a secret from its citizens and punishing people. This is terrible. It truly is a shameful country.

I, despite everything, will not be daunted by the difficulties. We only have one life to live.

Studying the history and the facts, I will not be deceived or yield to threats. I wish to live by following my principles. I will stay as I am.

The responsibility of scientists

Albert Einstein was a scientist who fled from Nazi Germany. At the end of 1938, the German Otto Hahn discovered the phenomenon of nuclear fission, which paved the way for many scientists to manufacture an atomic bomb. In the eyes of Leo Szilard, also a refugee in the United States to escape Nazism, Nazi Germany had embarked on the realization of this bomb and, if they were successful, the world would collapse; therefore, it was determined that the US would succeed the atomic project first. He persuaded Einstein to write to the President Franklin Roosevelt to suggest the manufacturing of an atomic bomb be investigated. Roosevelt then launched the "Manhattan Project". Between 50,000 and 100,000 people, workers or scientists were involved with the development of the atomic bomb. This project continued several years supported by many outstanding scientists. The financial investment for this project was so huge that even the entire annual budget of Japan at the time could not cover the costs. Thus, the first atomic bomb was born, and it was used in the Trinity nuclear test.

This test took place in 1945 in the desert of New Mexico and was a great success. The information was telegraphed to Potsdam, a small town outside of Berlin where the leaders of the USA, the UK and the Soviet Union were at the meeting - Truman, Churchill and Stalin. The anecdote says that Truman had whispered to Churchill, seated next to him, *«the atomic bomb is finally ready. After the war, the world will be ours.»* It was 16 July 1945.

By early 1945, as the war situation was about to be settled, Szilard collected a petition among his fellow scientists for the abandonment of nuclear tests. Einstein also wrote a second letter to Roosevelt. However, the president died suddenly in April without receiving Einstein's letter. In May, Nazi Germany collapsed. The manufacture of the atomic bomb had already lost its meaning, but because a huge budget had been devoted to its development, it was out of the question to give up. This bomb was dropped on Hiroshima.

Upon hearing the news, Einstein, distressed, said, *«if I were to be born again, I would never be a scientist.»*

Science also gives birth to horrors, very far from the intentions of the scientists. This was a good example.

Rethinking future energy consumption

The sun, an eternal source of energy, but...

Now, what are the possibilities for a future energy strategy? I think that solar energy is the best possibility. Previously, I introduced the figure that shows the reserved energy on earth (Figure 18, page 123). The sun provides us with over 14 times energy every year, compared to the ultimate reserves of carbon, natural gases, petrol, oil, shale, tar, sand, and uranium. The sun has been providing us with a massive amount of energy, which is incomparable to the resources the earth has been providing during 46 billion years. It is no longer the time to fantasise about nuclear energy. We should abandon nuclear energy as soon as possible and turn our attention to the effective use of solar energy without destroying our environment.

Yet, solar energy should not be used without limitations. Figure 23 shows the result of how Japan has been using energy.

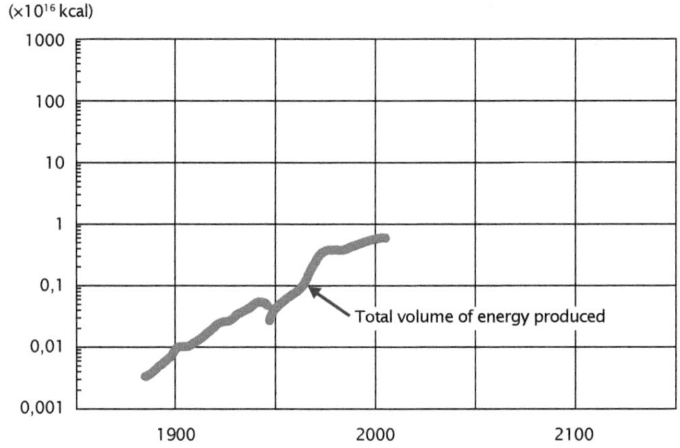

Source: Based on the EMDC Guide 2003 statistics on energy and the economy of Japan's Energy Data and Modelling Centre, Institute of Energy Economics (IEEJ) and general energy statistics agency for natural resources and energy of the Ministry of Economy, Trade and Industry

Figure 23: The result of how Japan has been using energy

The horizontal axis represents the years. The vertical axis represents the quantity of energy, and going up one cell means 10 times more. The thick grey line represents the actual values that start from 1880.

In the era of Meiji 13 (1880), when Japan finally started to form a modern society, Japan entered into the period to promote high-speed modernisation, with a slogan to catch up and overtake the USA and Europe, by using enormous amounts of energy.

While using huge amounts of energy, Japan entered the Pacific War before 1950. The use of energy decreased during the war, but it restarted again after the war with a period of high economic growth. Then the bubble economy arrived followed by the bubble burst, and since then it has been stable.

I drew the dotted lines just below 100 on the vertical axis (Figure 24, page 156). This is the quantity of energy the sun provides during one year to our land of 378,000 square kilometres. That is at 100%, and one cell below is at 10%, one cell further down is at 1%. Further below, there is a line, which indicates 0.2%. That is the separate amount of energy in the total solar energy, which is consumed to generate natural phenomena such as wind, waves and atmospheric current.

We are now at just over 2,000. If we adapt our consumption of energy to the energy provided by the sun, we would use 0.6% of it. We artificially consume three times more energy than that which solar energy produces for natural phenomena, in order to maintain our luxury lifestyles. We should be grateful for nature as it is still resisting despite our wasteful abuse of its resources.

What happens if we persist with this mentality of *"the more we consume energy, the more prosperous we become?"* I demonstrate the possible estimation with the dotted arrows. (Figure 24, page 156)

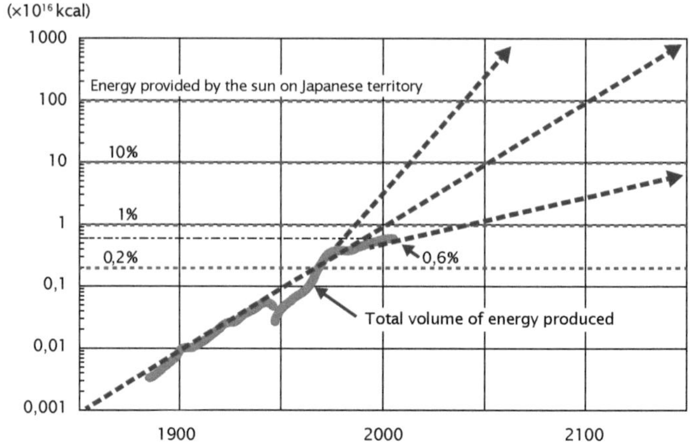

Note: Calculation based on a solar constant of 1,382 kW/m², an average sunshine in Japan of 256 kcal/cm² and an area of 378,000 km²

Source: Drawn from the EMDC 2003 energy statistics guide and economy of Japan's Energy Data and Modelling Centre, Institute of Energy Economics (IEEJ) and general energy statistics of the agency for natural resources and energy of the Ministry of Economy, Trade and Industry

Figure 24: The problematic environmental situation of Japan

If we follow the central arrow, by the time it reaches the 2050s from the present (the next cell), humans would want to artificially use 10% of solar energy. In another 50 years, it will go up to 100%. We would want to use the maximum energy the sun can provide. Furthermore, in 2150, which is in another 50 years, we would want to use 10 times more energy than the sun can provide. Can we really keep living like this?

What does it mean to "live considerately"?

However, I don't think the reality will turn out like my estimation. The Ministry of Economy, Trade and Industry (formally called the Ministry of International Trade and Industry when Figure 28 was made), shows that the amount of energy consumed by an individual Japanese person will increase (Figure 25), whereas the population of Japan will decrease.

Up until 2030, 0.5% of the population will decrease every year. After that 0.9% will decrease until 2050, after which the population will decrease 1%

each year. That is the Government's estimation. Therefore, it is possible to create a society where we don't consume as much energy, and it is necessary to do so. Along with the population decrease, each individual will reduce the use of energy by just 1%. That is not too difficult to do. If we can do it, by the time we have passed 2050, we will be able to approach 0.2% of the consumption of energy, which is equivalent to the solar energy consumed to cause natural phenomena. This estimate indicates that Japan can become such a society. It is the minimum goal we should pursue.

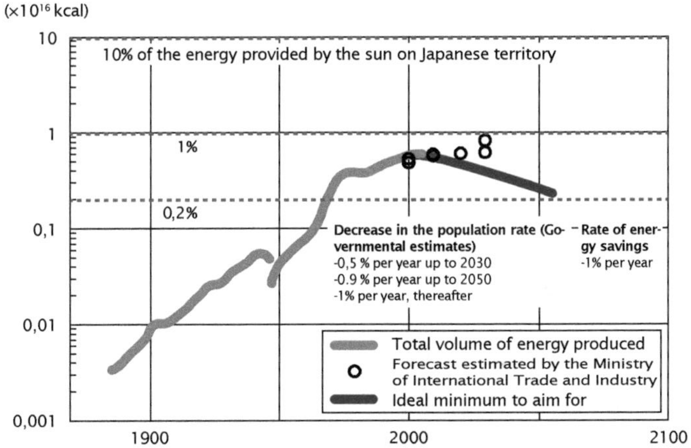

Source: Developed from the EMDC Guide 2003 statistics on energy and the economy of Japan's Energy Data and Modelling Centre, Institute of Energy Economics (IEEJ) and general energy statistics agency for natural resources and energy of the Ministry of Economy, Trade and Industry. "Forecasts from the Ministry of International Trade and Industry" was reported in June 1884, based on the report (revised in July 2001) of the Supply Group of Energy Investigation Committee (data for 2000 and 2010), long-term forecasts of IEEJ (Data for 2010 and 2020) and the report of the vision for the 21st Century Energy Committee of the Ministry of Economy, Trade, and Industry published in October 1986 (data for 2030)

Figure 25: Future options

The blue earth, floating in the pitch-black universe, the lights in Japan looked down on from space is as bright as daylight. Even without nuclear power plants, we have enough electricity. But, ultimately, it's not about whether we have sufficient energy or not.

Nuclear energy destroys DNAs, the chains to unite our lives. Nuclear power plants evoke discrimination and always force the most vulnerable people of societies into oppression. Nuclear energy must not be used. All nuclear power plants must be abandoned immediately.

At the same time, I realise that, as a human born onto this exceptional planet, the only planet where life has prospered in such richness, I should make the most effort to protect the ecosystem, of which I am also a part.

I am not saying living an affluent life is a bad thing. By pursuing an affluent life, we can shift our consciousness to be able to live happily without consuming so much energy, and changing the social constructions little by little to less energy consuming systems.

The American writer, Raymond Chandler, in his novel "Playback", wrote as follows: *"We have to be strong in order to live. We have no right to live if we are not considerate".* In addition to "we have to be strong in order to live", he insisted we be "considerate". What does it mean to live considerately? I have been asking myself this question for a long time. Recently, I have come to realise that consideration is defined by how I feel about other living beings that have more difficulties than I do, about all children in the world, including Japan, cows and horses, and all other living beings; always standing up for the weaker people in society; and living considerately. That is the most important, and the truest strength. Keeping this faith in mind, I will continue on my quest.

Four years have passed...

Friday, 30 August 2013, for the first time I participated as a member of the drum band, the anti-nuclear protest which has been repeated every Friday evening in front of the Prime Minister's residence. I have wanted to do it for a long time, but my main field of battle is the nuclear energy research in which I best fight.

For over forty years I have worked to end the nuclear power plants, but against the State authority, the combined influences of industry giants and the power of media collusion with them, my forces have been virtually inoperative. I always wished to stop nuclear power plants before a catastrophic accident occurred, but unfortunately, the Fukushima accident occurred. I wondered to myself, *"What was my life?"* I feel hopeless. While giving lectures to people all over Japan, I kept thinking that *"I would prefer not to have any listeners, I would rather the accident had not happened."*

When I participated in the protest meeting, I felt that there are many people who are committed to never forgetting about the Fukushima accident. I was pleased to see those people. If there are still people like them, I may be able to continue working to stop nuclear energy.

Formerly, Chaplin said as the protagonist in Monsieur Verdoux, the film he created after the Second World War: *«If you kill one person, you are a murderer. But if you kill a million, you are a hero.»*

History has shown that crimes committed by those in power can only be sentenced by a higher power. The day crimes of power are condemned at the hands of the people seems far away. But sharing the heartrending cry of the victims, I intend to continue the fight without giving up.

This book summarises the content of my public appearances over the past year, which I have updated with further facts, which subsequently came to my knowledge. Great thanks to Haruko Mine, Gentosha Renaissance who helped me write and Yumi Sato for editing. I would be happy for this work to contribute to stopping the mad race of the nuclear industry and I hope that the victims, who are still suffering from the difficulties, manage to escape from the current situation.

June 2015

Hiroaki KOIDÉ, born in Tokyo in 1949. Obtained a Nuclear Engineering Degree at the Faculty of Engineering, Tohoku University in 1972. Nuclear engineer at Kyoto University Research Reactor Institute (KURRI), assistant professor, specialized in radiation measurement and nuclear safety. He testified on the insufficiency of nuclear safety at the Ikata nuclear power plant litigation as one of the plaintiffs' (residents') witnesses. He authored numerous books, such as «Beyond the reality of radioactive contamination,» Hokuto publication, 1992 (republicated by Kawade Shobo, 2011), «Hidden Truth: Nuclear Power = Nuclear Arms» Soshisha, 2010, «Lies About Nuclear Power Plant» Fuyo-sha, 2011.
Advisor to the Japanese government and its Prime Minister Naoto KAN at the symposium in 2013, New York, about the medical and ecological consequences of the Fukushima accident. Some further participations: 2011 University of California, Berkeley, USA; 2012 University of Chicago, USA: *Atomic Age Symposium;* 2014 *The 4th Citizen-Scientist International Symposium on Radiation Protection, Tokyo.*

Taeko TOMIYAMA, Japanese painter (born 1920). Her creations explore in an imaginative way the entire world, both East and West, in its historical contents (war, colonialism, catastrophes). Her visual art is the expression of her poetic vision, of her memory, and her social responsibility, asking the public to remember and make the link to our turbulent world today.
Immediately after the triple disasters (earthquake, tsunami and nuclear power meltdown) of Fukushima March 11, 2011 she started creating a series of five paintings «Revelations from the sea», her response to the disaster, from which n° 3 and 4 where chosen to illustrate this book's topic.
Taeko TOMIYAMA, in collaboration with the musician Yuji TALAHASHI, created forceful audio visual works.

(based on an article by Rebecca JENNISON, teaching at the Kyoto Seika University.)

n°3 "FUKUSHIMA: Spring of Caesium-137"; Oil on canvas 1120 x 1620

What at first glance appears to be a painting of a commonly seen celebratory theme, Japan in springtime, comes to have foreboding associations. Here radioactive particles invade the spring scene visible to the eye. The dark, demon-like god rides on the wind, scattering particles of Caesium-137 along with seasonal cherry blossoms over land and sea.

n°4 "JAPAN: Nuclear Power Plant"; Oil on canvas 970 x 1455

This image of the bleak, crumbling frame of the plant, a skeleton against an equally bleak landscape, speaks more boldly of the fragility and hubris of this man-made accident than photos or words. It speaks of too much faith in technology and the thirst of a "triumphant" economy. Again the artiste urges us to learn from our past errors. This message is for all of us, given the obvious links between nuclear arms and nuclear energy.

Publications by AnimaViva multilingüe

Golden Nihon Collection

director Jacques KERIGUY

IKKYÛ. L'impertinence au service de la foi

Sôshô YAMADA, 2012

printed book ISBN 978-99920-1-966-5 (bilingual French/Japanese)

2th edition

printed book ISBN 978-99920-68-18-2 (bilingual French/Japanese)
Epub: ISBN: 978-99920-68-19-9; Mobi: ISBN: 978-99920-68-20-5
(in preparation: single French and single Japanese versions)

Penser le nucléaire. Autopsie d'une illusion

Hiroaki KOIDÉ, 2015

Afterword Jean-Jacques DELFOUR/René de CECCATTY
printed book French ISBN 978-99920-68-06-9
Epub: ISBN: 978-99920-68-16-8; Mobi: ISBN: 978-99920-68-17-5
(English in preparation: *Rethinking nuclear energy, autopsie of an illusion*)

L'Étrange voyage de Confucius au Japon

Olivier ANSART, 2015

printed book French ISBN: 978-99920-68-07-6
Epub: ISBN: 978-99920-68-08-3; Mobi: ISBN: 978-99920-68-09-0
(English 2015: *The strange travel of Confucius to Japan*)

MINAKATA Kumagusu. L'émergence d'une pensée écologique entre Orient et Occident

Satoshi OHARA, Jean-Christophe VALMALETTE, 2015

printed book French ISBN 978-99920-68-21-2 (English in preparation)
Epub: ISBN: 978-99920-68-22-9; Mobi: ISBN: 978-99920-68-23-6

In preparation:

Gidayû TAKEMOTO. Voix de la poupée

Akiko ÔTA, 2016

printed book, e-books (French)

Golden Tagore Collection

published under the Memorandum of Understanding (MoU) between *Visva-Bharati University*, Santiniketan (India) and *AnimaViva multilingüe*, Principality of Andorra

The Original Gitanjali by Rabindranath Tagore

Rabindranath TAGORE / Udaya Narayan SINGH (narration, analysis, translation), 2014

ePub: ISBN 978-99920-68-25-0 ; mobiKindle: ISBN 978-99920-68-04-5
printed book, 2015: ISBN 978-99920-68-05-2
(English, Spanish, Catalan, French)

In preparation:

Beyond Language: Towards Silence
Tagore's Journey into Art & Poetry
Udaya Narayana SINGH, 2015
Printed book, e-books (English, French, Spanish, Catalan)

Series Essais

Ecriture et Plasticité de Pensée
Marc-Williams DEBONO, 2013
Forword Michel CAZENAVE
ePub: ISBN 978-99920-68-19-9; mobiKindle: ISBN 978-99920-68-28-1
printed book 2015 (French)

In preparation:

L'essence de l'amour selon Søren KIERKEGAARD
Francesc TORRALBA, 2015
printed book, e-books, (French, English, Spanish, Catalan)

DALAI LAMA et Carl Friedrich von WEIZSÄCKER
Where Spirituality & Science meet, 2015
printed book, e-books, (English, French, Tibetan)

Series Poetry

Eloge de l'impossible, Angel IYÉ, 2015
printed book French ISBN: 978-99920-68-24-3
Epub: ISBN: 978-99920-68-25-0; Mobi: ISBN: 978-99920-68-26-7
printed book Catalan ISBN: 978-99920-68-27-4
Epub: ISBN: 978-99920-68-28-1; Mobi: ISBN: 978-99920-68-29-8
printed book bilingual French/Catalan ISBN: 978-99920-68-30-4
Epub: ISBN: 978-99920-68-31-1; Mobi: ISBN: 978-99920-68-32-8

Series Accents
director Heinz WISMANN

In preparation:

Avant la philosophie, Arnaud VILLANI, 2015
printed book, e-books, (French)

Penser le design, Jean Philippe TESTEFORT, 2015
printed book, e-books, (French)

L'éthique des l'éthiques, Heinz WISMANN, 2015
printed book, e-books, (French, English, German)

Devant la mort, Lise HADDAD, 2015
printed book, e-books, (French)

Golden Gibran Collection

In preparation:

Le Prophète, Kahlil GIBRAN, 2015
printed book, e-books, (French, English, Arab, Spanish, Catalan)